LECTURES ON THE h-COBORDISM THEOREM

BY

JOHN MILNOR

NOTES BY

L. SIEBENMANN

AND

J. SONDOW

PRINCETON, NEW JERSEY
PRINCETON UNIVERSITY PRESS
1965

Printed and bound by CPI Group (UK) Ltd, Croydon, CR0 4YY

§0. Introduction

These are notes for lectures of John Milnor that were given as a seminar on differential topology in October and November, 1963 at Princeton University.

Let W be a compact smooth manifold having two boundary components V and V' such that V and V' are both deformation retracts of W. Then W is said to be a <u>h-cobordism</u> between V and V'. The <u>h-cobordism theorem</u> states that if in addition V and (hence) V' are simply connected and of dimension greater than 4, then W is diffeomorphic to $V \times [0, 1]$ and (consequently) V is diffeomorphic to V'. The proof is due to Stephen Smale [6]. This theorem has numerous important applications — including the proof of the generalized Poincaré conjecture in dimensions > 4 — and several of these appear in §9. Our main task, however, is to describe in some detail a proof of the theorem.

Here is a very rough outline of the proof. We begin by constructing a Morse function for W (§2.1), i.e. a smooth function $f : W \longrightarrow [0, 1]$ with $V = f^{-1}(0)$, $V' = f^{-1}(1)$ such that f has finitely many critical points, all nondegenerate and in the interior of W. The proof is inspired by the observation (§3.4) that W is diffeomorphic to $V \times [0, 1]$ if (and only if) W admits a Morse function as above with no critical points. Thus in §§4-8 we show that under the hypothesis of the theorem it is possible to simplify a given Morse function

f until finally all critical points are eliminated. In §4, f is adjusted so that the level $f(p)$ of a critical point p is an increasing function of its index. In §5, geometrical conditions are given under which a pair of critical points p, q of index λ and $\lambda + 1$ can be eliminated or 'cancelled'. In §6, the geometrical conditions of §5 are replaced by more algebraic conditions —— given a hypothesis of simple connectivity. In §8, the result of §5 allows us to eliminate all critical points of index 0 or n, and then to replace the critical points of index 1 and $n - 1$ by equal numbers of critical points of index 3 and $n - 3$, respectively. In §7 it is shown that the critical points of the same index λ can be rearranged among themselves for $2 \leq \lambda \leq n - 2$ (§7.6) in such a way that all critical points can then be cancelled in pairs by repeated application of the result of §6. This completes the proof.

Two acknowledgements are in order. In §5 our argument is inspired by recent ideas of M. Morse [11][32] which involve alteration of a gradient-like vector field for f, rather than by the original proof of Smale which involves his 'handlebodies'. We in fact never explicitly mention handles or handlebodies in these notes. In §6 we have incorporated an improvement appearing in the thesis of Dennis Barden [33], namely the argument on our pages 72-73 for Theorem 6.4 in the case $\lambda = 2$, and the statement of Theorem 6.6 in the case $r = 2$.

The h-cobordism theorem can be generalized in several directions. No one has succeeded in removing the restriction that V and V' have dimension > 4. (See page 113.) If we omit the restriction that V and (hence) V' be simply connected, the theorem becomes false. (See Milnor [34].) But it will remain true if we at the same time assume that the inclusion of V (or V') into W is a simple homotopy equivalence in the sense of J. H. C. Whitehead. This generalization, called the s-cobordism theorem, is due to Mazur [35], Barden [33] and Stallings. For this and further generalizations see especially Wall [36]. Lastly, we remark that analogous h- and s-cobordism theorems hold for piecewise linear manifolds.

Contents

Section 1. The Cobordism Category

First some familiar definitions. Euclidean space will be denoted by $R^n = \{(x_1,\ldots,x_n) | x_i \in R, \ i = 1,\ldots,n\}$ where R = the real numbers, and Euclidean half-space by $R^n_+ = \{(x_1,\ldots,x_n) \in R^n | x_n \geq 0\}$.

<u>Definition 1.1.</u> If V is any subset of R^n, a map $f\colon V \longrightarrow R^m$ is <u>smooth</u> or <u>differentiable of class C^∞</u> if f can be extended to a map $g\colon U \longrightarrow R^m$, where $U \supset V$ is open in R^n, such that the partial derivatives of g of all orders exist and are continuous.

<u>Definition 1.2.</u> A <u>smooth n-manifold</u> is a topological manifold W with a countable basis together with a <u>smoothness structure</u> $\mathcal{S}$ on M. $\mathcal{S}$ is a collection of pairs (U,h) satisfying four conditions:

(1) Each $(U,h) \in \mathcal{S}$ consists of an open set $U \subset W$ (called a <u>coordinate neighborhood</u>) together with a homeomorphism h which maps U onto an open subset of either R^n or R^n_+.

(2) The coordinate neighborhoods in $\mathcal{S}$ cover W.

(3) If (U_1,h_1) and (U_2,h_2) belong to $\mathcal{S}$, then

$$h_1 h_2^{-1}\colon h_2(U_1 \cap U_2) \longrightarrow R^n \text{ or } R^n_+$$

is smooth.

(4) The collection $\mathcal{S}$ is maximal with respect to property (3); i.e. if any pair (U, h) not in $\mathcal{S}$ is adjoined to $\mathcal{S}$, then property (3) fails.

The boundary of W, denoted Bd W, is the set of all points in W which do not have neighborhoods homeomorphic to R^n (see Munkres [5, p.8]).

Definition 1.3. $(W; V_0, V_1)$ is a smooth manifold triad if W is a compact smooth n-manifold and Bd W is the disjoint union of two open and closed submanifolds V_0 and V_1 .

If $(W; V_0, V_1)$, $(W'; V_1', V_2')$ are two smooth manifold triads and $h: V_1 \longrightarrow V_1'$ is a diffeomorphism (i.e. a homeomorphism such that h and h^{-1} are smooth), then we can form a third triad $(W \cup_h W'; V_0, V_2')$ where $W \cup_h W$ is the space formed from W and W' by identifying points of V_1 and V_1' under h, according to the following theorem.

Theorem 1.4. There exists a smoothness structure $\mathcal{S}$ for $W \cup_h W'$ compatible with the given structures (i.e. so that each inclusion map $W \longrightarrow W \cup_h W'$, $W' \longrightarrow W \cup_h W'$ is a diffeomorphism onto its image.)

$\mathcal{S}$ is unique up to a diffeomorphism leaving V_0, $h(V_1) = V_1'$, and V_2' fixed.

The proof will be given in § 3 .

Definition 1.5. Given two closed smooth n-manifolds M_0 and M_1 (i.e. M_0 , M_1 compact, Bd M_0 = Bd $M_1 = \emptyset$), a cobordism from M_0 to M_1 is a 5-tuple, $(W; V_0, V_1; h_0, h_1)$, where $(W; V_0, V_1)$ is a smooth manifold triad and $h_i: V_i \longrightarrow M_i$ is a diffeomorphism, i = 0, 1. Two cobordisms $(W; V_0, V_1; h_0, h_1)$ and $(W'; V_0', V_1'; h_0', h_1')$ from M_0 to M_1 are equivalent if there exists a diffeomorphism $g: W \longrightarrow W'$ carrying V_0 to V_0' and V_1 to

V_i' such that for $i = 0,1$ the following triangle commutes:

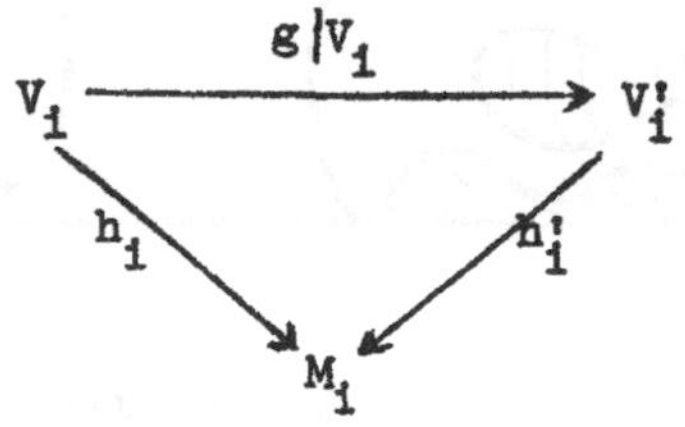

Then we have a category (see Eilenberg and Steenrod, [2,p.108]) whose objects are closed manifolds and whose morphisms are equivalence classes c of cobordisms. This means that cobordisms satisfy the following two conditions. They follow easily from 1.4 and 3.5, respectively.

(1) Given cobordism equivalence classes c from M_0 to M_1 and c' from M_1 to M_2, there is a well-defined class cc' from M_0 to M_2. This composition operation is associative.

(2) For every closed manifold M there is the identity cobordism class ι_M = the equivalence class of $(M \times I; M \times 0, M \times 1; p_0, p_1)$, $p_i(x,i) = x$, $x \in M$, $i = 0,1$. That is, if c is a cobordism class from M_1 to M_2, then

$$\iota_{M_1} c = c = c\,\iota_{M_2} .$$

Notice that it is possible that $cc' = \iota_M$, but c is not ι_M. For example

FIGURE 1

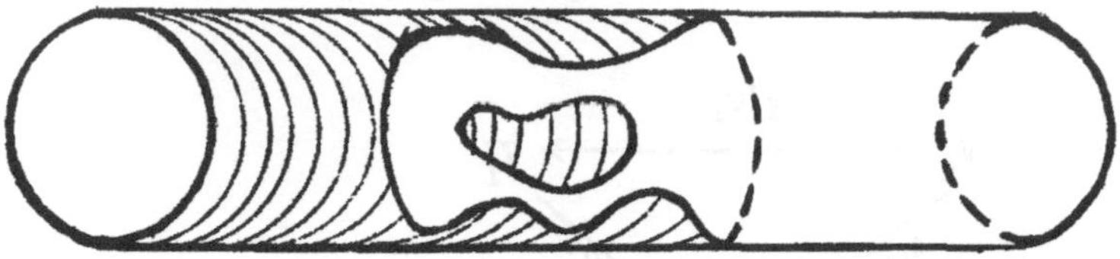

c is shaded. c' is unshaded.

Here c has a right inverse c', but no left inverse. Note that the manifolds in a cobordism are not assumed connected.

Consider cobordism classes from M to itself, M fixed. These form a monoid H_M, i.e. a set with an associative composition with an identity. The invertible cobordisms in H_M form a group G_M. We can construct some elements of G_M by taking $M = M'$ below.

Given a diffeomorphism $h: M \longrightarrow M'$, define c_h as the class of $(M \times I; M \times 0, M \times 1; j, h_1)$ where $j(x,0) = x$ and $h_1(x,1) = h(x)$, $x \in M$.

Theorem 1.6. $c_h c_{h'} = c_{h'h}$ for any two diffeomorphisms $h: M \longrightarrow M'$ and $h': M' \longrightarrow M''$.

Proof: Let $W = M \times I \cup_h M' \times I$ and let $j_h: M \times I \longrightarrow W$, $j_{h'}: M' \times I \longrightarrow W$ be the inclusion maps in the definition of $c_h c_{h'}$. Define $g: M \times I \longrightarrow W$ as follows:

$$g(x,t) = j_h(x,2t) \qquad 0 \leq t \leq \frac{1}{2}$$

$$g(x,t) = j_{h'}(h(x),2t-1) \qquad \frac{1}{2} \leq t \leq 1 .$$

Then g is well-defined and is the required equivalence.

Definition 1.7. Two diffeomorphisms $h_0, h_1: M \longrightarrow M'$ are (smoothly) isotopic if there exists a map $f: M \times I \longrightarrow M'$ such that

(1) f is smooth,

(2) each f_t, defined by $f_t(x) = f(x,t)$, is a diffeomorphism,

(3) $f_0 = h_0$, $f_1 = h_1$.

Two diffeomorphisms $h_0, h_1: M \longrightarrow M'$ are pseudo-isotopic* if there is a diffeomorphism $g: M \times I \longrightarrow M' \times I$ such that $g(x,0) = (h_0(x),0)$, $g(x,1) = (h_1(x),1)$.

Lemma 1.8. Isotopy and pseudo-isotopy are equivalence relations.

Proof: Symmetry and reflexivity are clear. To show transitivity, let $h_0, h_1, h_2: M \longrightarrow M'$ be diffeomorphisms and assume we are given isotopies $f, g: M \times I \longrightarrow M'$ between h_0 and h_1 and between h_1 and h_2 respectively. Let $m: I \longrightarrow I$ be a smooth monotonic function such that $m(t) = 0$ for $0 \leq t \leq 1/3$, and $m(t) = 1$ for $2/3 \leq t \leq 1$. The required isotopy $k: M \times I \longrightarrow M'$ between h_0 and h_1 is now defined by $k(x,t) = f(x,m(2t))$ for $0 \leq t \leq 1/2$, and $k(x,t) = g(x,m(2t-1))$ for $1/2 \leq t \leq 1$. The proof of transitivity for pseudo-isotopies is more difficult and follows from Lemma 6.1 of Munkres [5,p.59].

---

* In Munkres' terminology h_0 is "I-cobordant" to h_1. (See [5,p.62].) In Hirsch's terminology h_0 is "concordant" to h_1.

It is clear that if h_0 and h_1 are isotopic then they are pseudo-isotopic, for if $f: M \times \dot{I} \longrightarrow M'$ is the isotopy, then $\hat{f}: M \times I \longrightarrow M' \times I$, defined by $\hat{f}(x,t) = (f_t(x),t)$, is a diffeomorphism, as follows from the inverse function theorem, and hence is a pseudo-isotopy between h_0 and h_1. (The converse for $M = S^n$, $n \geq 8$ is proved by J. Cerf [39].) It follows from this remark and from 1.9 below that if h_0 and h_1 are isotopic, then $c_{h_0} = c_{h_1}$.

<u>Theorem 1.9</u>. $c_{h_0} = c_{h_1} \Longleftrightarrow$ h_0 <u>is</u> <u>pseudo-isotopic</u> <u>to</u> h_1

<u>Proof</u>: Let $g: M \times I \longrightarrow M' \times I$ be a pseudo-isotopy between h_0 and h_1. Define $h_0^{-1} \times I: M' \times I \longrightarrow M \times I$ by $(h_0^{-1} \times 1)(x,t) = (h_0^{-1}(x),t)$. Then $(h_0^{-1} \times 1) \circ g$ is an equivalence between c_{h_1} and c_{h_0}.

The converse is similar.

Section 2. Morse Functions

We would like to be able to factor a given cobordism into a composition of simpler cobordisms. (For example the triad in Figure 2 can be factored as in Figure 3.) We make this notion precise in what follows.

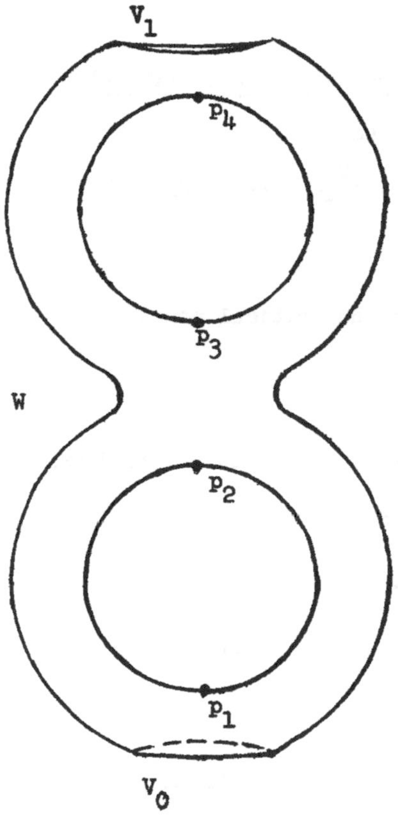

FIGURE 2

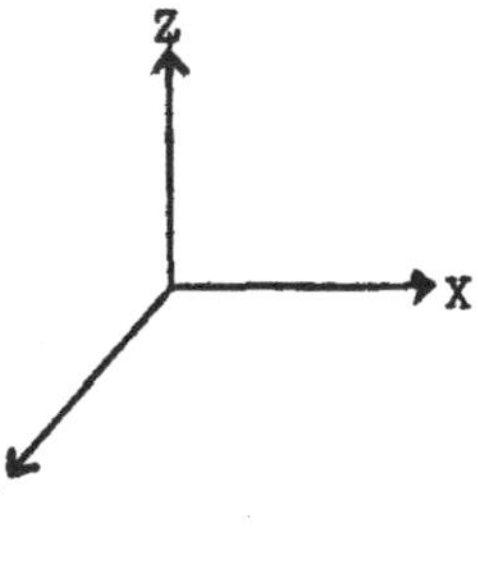

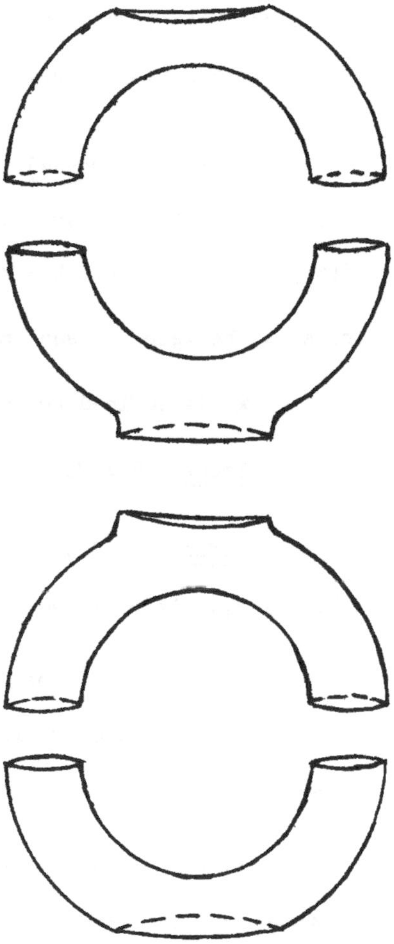

FIGURE 3

Definition 2.1. Let W be a smooth manifold and $f: W \longrightarrow R$ a smooth function. A point $p \in W$ is a critical point of f if, in some coordinate system, $\left.\frac{\partial f}{\partial x_1}\right|_p = \left.\frac{\partial f}{\partial x_2}\right|_p = \ldots = \left.\frac{\partial f}{\partial x_n}\right|_p = 0$. Such a point is a non-degenerate critical point if $\det\left(\left.\frac{\partial^2 f}{\partial x_i \partial x_j}\right|_p\right) \neq 0$. For example, if in Figure 2 f is the height function (projection into the z-axis), then f has four critical points p_1, p_2, p_3, p_4, all non-degenerate.

Lemma 2.2 (Morse). If p is a non-degenerate critical point of f, then in some coordinate system about p, $f(x_1,\ldots,x_n) = \text{constant} - x_1^2 - \ldots - x_\lambda^2 + x_{\lambda+1}^2 + \ldots + x_n^2$ for some λ between 0 and n.

λ is defined to be the index of the critical point p.

Proof: See Milnor [4, p.6] .

Definition 2.3. A Morse function on a smooth manifold triad $(W; V_0, V_1)$ is a smooth function $f: W \longrightarrow [a,b]$ such that

(1) $f^{-1}(a) = V_0$, $f^{-1}(b) = V_1$,

(2) All the critical points of f are interior (lie in $W - \text{Bd } W$) and are non-degenerate.

As a consequence of the Morse Lemma, the critical points of a Morse function are isolated. Since W is compact, there are only finitely many of them.

<u>Definition 2.4.</u> The <u>Morse number</u> μ of $(W; V_0, V_1)$ is the minimum over all Morse functions f of the number of critical points of f.

This definition is meaningful in view of the following existence theorem.

<u>Theorem 2.5. Every smooth manifold triad $(W; V_0, V_1)$ possesses a Morse function.</u>

The proof will occupy the next 8 pages.

<u>Lemma 2.6. There exists a smooth function $f: W \longrightarrow [0,1]$ with $f^{-1}(0) = V_0$, $f^{-1}(1) = V_1$, such that f has no critical point in a neighborhood of the boundary of W.</u>

<u>Proof:</u> Let $U_1, \ldots, U_k$ be a cover of W by coordinate neighborhoods. We may assume that no U_i meets both V_0 and V_1, and that if U_i meets Bd W the coordinate map $h_i: U_i \longrightarrow R^n_+$ carries U_i onto the intersection of the open unit ball with R^n_+.

On each set U_i define a map

$$f_i: U_i \longrightarrow [0,1]$$

as follows. If U_i meets V_0, [respectively V_1] let $f_i = Lh_i$ where L is the map

$$L\vec{x} = x_n \quad [\text{respectively } 1 - x_n] \, .$$

If U_i does not meet Bd W, put $f_i = 1/2$ identically.

Choose a partition of unity $\{\varphi_i\}$ subordinate to the cover $\{U_i\}$ (see Munkres [5,p.18]) and define a map $f\colon W \longrightarrow [0,1]$ by

$$f(p) = \varphi_1(p)f_1(p) + \dots + \varphi_k(p)f_k(p)$$

where $f_i(p)$ is understood to have the value 0 outside U_i. Then f is clearly a well defined smooth map to $[0,1]$ with $f^{-1}(0) = V_0$, $f^{-1}(1) = V_1$. Finally we verify that $df \neq 0$ on Bd W. Suppose $q \in V_0$ [respectively $q \in V_1$]. Then, for some i, $\varphi_i(q) > 0$, and $q \in U_i$. Let $h_i(p) = (x^1(p),\dots,x^n(p))$. Then

$$\frac{\partial f}{\partial x^n} = \sum_{j=1}^{k} f_j \frac{\partial \varphi_j}{\partial x^n} + \left\{\varphi_1 \frac{\partial f_1}{\partial x^n} + \dots + \varphi_i \frac{\partial f_i}{\partial x^n} + \dots\right\}.$$

Now $f_j(z)$ has the same value, 0, [respectively 1] for all j and $\sum_{j=1}^{k} \frac{\partial \varphi_j}{\partial x^n} = \frac{\partial}{\partial x^n}\{\sum_{j=1}^{k} \varphi_j\} = 0$. So, at q, the first summand is zero. The derivative $\frac{\partial f_i}{\partial x^n}(q)$ equals 1 [respectively -1] and it is easily seen that the derivatives $\frac{\partial f_j}{\partial x^n}(q)$ all have the same sign as $\frac{\partial f_i}{\partial x^n}(q)$, $j = 1,\dots,k$. Thus $\frac{\partial f}{\partial x^n}(q) \neq 0$. It follows that $df \neq 0$ on Bd W, and hence $df \neq 0$ in a neighborhood of Bd W.

The remainder of the proof is more difficult. We will alter f by stages in the interior of W eliminating the degenerate critical points. To do this we need three lemmas which apply to Euclidean space.

<u>Lemma A (M. Morse). If f is a C^2 mapping of an open subset $U \subset R^n$ to the real line, then, for almost all linear mappings $L: R^n \longrightarrow R$, the function $f + L$ has only nondegenerate critical points.</u>

By "almost all" we mean except for a set which has measure zero in $\mathrm{Hom}_R(R^n,R) \cong R^n$.

<u>Proof:</u> Consider the manifold $U \times \mathrm{Hom}_R(R^n,R)$. It has a submanifold $M = \{(x,L) | d(f(x) + L(x)) = 0\}$. Since $d(f(x) + L(x)) = 0$ means that $L = -df(x)$ it is clear that the correspondence $x \longrightarrow (x, -df(x))$ is a diffeomorphism of U onto M. Each $(x,L) \in M$ corresponds to a critical point of $f + L$, and this critical point is degenerate precisely when the matrix $(\frac{\partial^2 f}{\partial x_i \partial x_j})$ is singular. Now we have a projection $\pi: M \longrightarrow \mathrm{Hom}(R^n,R)$ sending (x,L) to L. Since $L = -df(x)$, the projection is nothing but $x \longrightarrow -df(x)$. Thus π is critical at $(x,L) \in M$ precisely when the matrix $d\pi = -(\partial^2 f/\partial x_i \partial x_j)$ is singular. It follows that $f + L$ has a degenerate critical point (for some x) if and only if L is the image of a critical point of $\pi: M \longrightarrow \mathrm{Hom}_R(R^n,R) \cong R^n$. But, by the theorem of Sard (see de Rham [1,p.10]):

<u>If $\pi: M^n \longrightarrow R^n$ is any C^1 map, the image of the set of critical points of π has measure zero in R^n.</u>

This gives the desired conclusion.

Lemma B. *Let K be a compact subset of an open set U in R^n. If $f\colon U \longrightarrow R$ is C^2 and has only nondegenerate critical points in K, then there is a number $\delta > 0$ such that if $g\colon U \longrightarrow R$ is C^2 and at all points of K satisfies*

$$(1)\quad \left|\frac{\partial f}{\partial x_i} - \frac{\partial g}{\partial x_i}\right| < \delta\,, \qquad (2)\quad \left|\frac{\partial^2 f}{\partial x_i \partial x_j} - \frac{\partial^2 g}{\partial x_i \partial x_j}\right| < \delta$$

$i,j = 1,\ldots,n$, *then g likewise has only nondegenerate critical points in K.*

Proof: Let $\quad |df| = [(\frac{\partial f}{\partial x_1})^2 + \ldots + (\frac{\partial f}{\partial x_n})^2]^{1/2}$.

Then $|df| + |\det(\frac{\partial^2 f}{\partial x_i \partial x_j})|$ is strictly positive on K. Let $\mu > 0$ be its minimum on K. Choose $\delta > 0$ so small that (1) implies that

$$||df| - |dg|| < \mu/2$$

and (2) implies that

$$\left||\det(\frac{\partial^2 f}{\partial x_i \partial x_j})| - |\det(\frac{\partial^2 g}{\partial x_i \partial x_j})|\right| < \mu/2 .$$

Then $|dg| + |\det(\frac{\partial^2 g}{\partial x_i \partial x_j})| > |df| + |\det(\frac{\partial^2 f}{\partial x_i \partial x_j})| - \mu/2 - \mu/2 \geq 0$

at all points in K. The result follows.

Lemma C. *Suppose $h\colon U \longrightarrow U'$ is a diffeomorphism of one open subset of R^n onto another and carries the compact set $K \subset U$ onto $K' \subset U'$. Given a number $\epsilon > 0$, there is a number $\delta > 0$ such that if a smooth map $f\colon U' \longrightarrow R$ satisfies*

$$|f| < \delta, \quad \left|\frac{\partial f}{\partial x_i}\right| < \delta, \quad \left|\frac{\partial^2 f}{\partial x_i \partial x_j}\right| < \delta \quad i, j = 1, \ldots, n$$

at all points of $K' \subset U'$, then $f \circ h$ satisfies

$$|f \circ h| < \epsilon, \quad \left|\frac{\partial f \circ h}{\partial x_i}\right| < \epsilon, \quad \left|\frac{\partial^2 f \circ h}{\partial x_i \partial x_j}\right| < \epsilon \quad i, j = 1, \ldots, n,$$

at all points of K.

Proof: Each of $f \circ h$, $\frac{\partial f \circ h}{\partial x_i}$, $\frac{\partial^2 f \circ h}{\partial x_i \partial x_j}$ is a polynomial function of the partial derivatives of f and of h from order 0 to order 2; and this polynomial vanishes when the derivatives of f vanish. But the derivatives of h are bounded on the compact set K. The result follows.

The C^2 topology on the set $F(M,R)$ of smooth real-valued functions on a compact manifold, M, (with boundary) may be defined as follows. Let $\{U_\alpha\}$ be a finite coordinate covering with coordinate maps $h_\alpha \colon U_\alpha \longrightarrow R^n$, and let $\{C_\alpha\}$ be a compact refinement of $\{U_\alpha\}$ (cf. Munkres [5, p.7]). For every positive constant $\delta > 0$, define a subset $N(\delta)$, of $F(M,R)$ consisting of all maps $g \colon M \longrightarrow R$ such that, for all α,

$$* \quad |g_\alpha| < \delta, \quad \left|\frac{\partial g_\alpha}{\partial x_i}\right| < \delta, \quad \left|\frac{\partial^2 g_\alpha}{\partial x_i \partial x_j}\right| < \delta$$

at all points in $h_\alpha(C_\alpha)$, where $g_\alpha = g h_\alpha^{-1}$ and $i, j = 1, \ldots, n$. If we take the sets $N(\delta)$ as a base of neighborhoods of the zero function in the additive group $F(M,R)$, the resulting topology is called the C^2 topology. The sets of the form $f + N(\delta) = N(f,\delta)$ give a base of neighborhoods of any map $f \in F(M,R)$, and $g \in N(f,\delta)$ means that, for all α,

$$|f_\alpha - g_\alpha| < \delta, \qquad \left|\frac{\partial f_\alpha}{\partial x_i} - \frac{\partial g_\alpha}{\partial x_i}\right| < \delta, \qquad \left|\frac{\partial^2 f_\alpha}{\partial x_i \partial x_j} - \frac{\partial^2 g_\alpha}{\partial x_i \partial x_j}\right| < \delta$$

at all points of $h_\alpha(C_\alpha)$.

It should be verified that the topology T we have constructed does not depend on the particular choice of coordinate covering and compact refinement. Let T' be another topology defined by the above procedure, and let primes denote things associated with this topology. It is sufficient to show that, given any set $N(\delta)$ in T, we can find a set $N'(\delta')$ in T' contained in $N(\delta)$. But this is an easy consequence of Lemma C.

We first consider a closed manifold M, i.e. a triad $(M, \emptyset, \emptyset)$, since this case is somewhat easier.

Theorem 2.7. *If M is a compact manifold without boundary, the Morse functions form an open dense subset of $F(M,R)$ in the C^2 topology.*

Proof: Let $(U_1,h_1),\ldots,(U_k,h_k)$ be a finite covering of M by coordinate neighborhoods. We can easily find compact sets $C_i \subset U_i$ such that $C_1, C_2 \ldots, C_k$ cover M.

We will say that f is "good" on a set $S \subset M$ if f has no degenerate critical points on S.

I) The set of Morse functions is open. For if $f: M \longrightarrow R$ is a Morse function, Lemma B says that, in a neighborhood N_i of f in $F(M, R)$, every function will be good in C_i. Thus, in the neighborhood $N = N_1 \cap \ldots \cap N_k$ of f, every function will be good in $C_1 \cup \ldots \cup C_k = M$.

II) The set of Morse functions is dense. Let N be a given neighborhood $f \in F(M, R)$. We improve f by stages. Let λ be a smooth function $M \longrightarrow [0,1]$ such that $\lambda = 1$ in a neighborhood of C_1 and $\lambda = 0$ in a neighborhood of $M - U_1$ For almost all choices of linear map $L: R^n \longrightarrow R$ the function $f_1(p) = f(p) + \lambda(p)\, L(h_1(p))$ will be good on $C_1 \subset U_1$ (Lemma A). We assert that if the coefficients of the linear map L are sufficiently small, then f_1 will lie in the given neighborhood N of f.

First note that f_1 differs from f only on a compact set $K = \text{Support } \lambda \subset U_1$. Setting $L(x) = L(x_1, \ldots, x_n) = \Sigma\, \ell_i x_i$, note that

$$f_1 h_1^{-1}(x) - f h_1^{-1}(x) = (\lambda h_1^{-1}(x))\, \Sigma\, \ell_i x_i$$

for all $x \in h_1(K)$. By choosing the ℓ_i sufficiently small we can clearly guarantee that this difference, together with its first and second derivatives, is less than any preassigned ϵ throughout the set $h_1(K)$. Now if ϵ is sufficiently small, then it follows from Lemma C that f_1 belongs to the neighborhood N.

We have obtained a function f_1 in N which is good on C_1. Applying Lemma B again, we can choose a neighborhood N_1 of f_1, $N_1 \subset N$, so that any function in N_1 is still good on C_1. This completes the first stage.

At the next stage, we simply repeat the whole process with f_1 and N_1, to obtain a function f_2 in N_1 good in C_2, and a neighborhood, N_2 of f_2, $N_2 \subset N_1$, such that any function in N_2 is still good on C_2. The function f_2 is automatically good on C_1 since it lies in N_1. Finally we obtain a function $f_k \in N_k \subset N_{k-1} \subset \cdots \subset N_1 \subset N$ which is good on $C_1 \cup \cdots \cup C_k = M$.

We are now in a position to prove

Theorem 2.5. On any triad (W, V_0, V_1), there exists a Morse function.

Proof: Lemma 2.6 provides a function $f: W \longrightarrow [0,1]$ such that (i) $f^{-1}(0) = V_0$, $f^{-1}(1) = V_1$

(ii) f has no critical points in a neighborhood of Bd W.

We want to eliminate the degenerate critical points in W - Bd W, always preserving the properties (i) and (ii) of f. Let U be an open neighborhood of Bd W on which f has no critical points. Because W is normal we can find an open neighborhood V of Bd W such that $\bar{V} \subset U$. Let $\{U_i\}$ be a finite cover of W by coordinate neighborhoods such that each set U_i lies in U or in $W - \bar{V}$. Take a compact refinement $\{C_i\}$ of $\{U_i\}$ and let C_0 be the union of all those C_i that lie in U. Just as for the closed manifold of the last theorem we can use Lemma B to show that in a sufficiently small neighborhood N of f, no function can have a degenerate critical point in C_0. Also f is bounded away from 0 and 1 on the compact set $W - V$.

Hence, in a neighborhood N' of f every function, g, satisfies the condition $0 < g < 1$ on $W - V$. Let $N_0 = N \cap N'$. We may suppose that the coordinate neighborhoods in $W - V$ are $U_1, \ldots, U_k$. From this point we proceed exactly as in the previous theorem. With the help of Lemma A we fnnd a function f_1 in N_0 which is good (i.e. has only nondegenerate critical points) on C_1, and a neighborhood N_1 of f_1, $N_1 \subset N_0$ in which every function is good in C_1. Repeating this process k times we produce a function $f_k \in N_k \subset N_{k-1} \subset \cdots \subset N_0$ which is good on $C_0 \cup C_1 \cup \ldots \cup C_k = M$. Since $f_k \subset N_0 \subset N'$ and $f_k|V = f|V$, f_k satisfies both conditions (i) and (ii). Hence f_k is a Morse function on (W, V_0, V_1).

<u>Remark</u>: It is not difficult to show that, in the C^2 topology, the Morse functions form an open dense subset of all smooth maps $f: (W, V_0, V_1) \longrightarrow ([0,1],0,1)$.

For some purposes it is convenient to have a Morse function in which no two critical points lie at the same level.

<u>Lemma 2.8.</u> *Let $f: W \longrightarrow [0,1]$ be a Morse function for the triad $(W; V_0, V_1)$ with critical points $p_1, \ldots, p_k$. Then f can be approximated by a Morse function g with the same critical points such that $g(p_i) \neq g(p_j)$ for $i \neq j$.*

<u>Proof</u>: Suppose that $f(p_1) = f(p_2)$. Construct a smooth function $\lambda: W \longrightarrow [0,1]$ such that $\lambda = 1$ in a neighborhood

U of p_1 and $\lambda = 0$ outside a larger neighborhood N, where $\bar{N} \subset W - \text{Bd}\, W$ and $\bar{N}$ contains no p_i for $i \neq 1$. Choose $\epsilon_1 > 0$ so small that $f_0 = f + \epsilon_1\lambda$ has values in $[0,1]$ and $f_0(p_1) \neq f_0(p_i)$, $i \neq 1$. Introduce a Riemannian metric for W (see Munkres [5, p.24]), and find c and c' so that $0 \leq c \leq |\text{grad } f|$ throughout the compact set $K = \text{closure } \{0 < \lambda < 1\}$ and $|\text{grad } \lambda| \leq c'$ on K. Let $0 < \epsilon < \min(\epsilon_1, c/c')$. Then $f_1 = f + \epsilon\lambda$ is again a Morse function, $f_1(p_1) \neq f(p_i)$ for $i \neq 1$, and f_1 has the same critical points as f. For on K,

$$\begin{aligned} |\text{grad } (f + \epsilon\lambda)| &\geq |\text{grad } f| - |\epsilon \text{ grad } \lambda| \\ &> c - \epsilon c' \\ &> 0 . \end{aligned}$$

And off K, $|\text{grad } \lambda| = 0$, so $|\text{grad } f_1| = |\text{grad } f|$. Continuing inductively, we obtain a Morse function g which separates all the critical points. This completes the proof.

Using Morse functions we can now express any "complicated" cobordism as a composition of "simpler" cobordisms.

Definition. Given a smooth function $f: W \longrightarrow R$, a critical value of f is the image of a critical point.

Lemma 2.9. *Let* $f: (W; V_0, V_1) \longrightarrow ([0,1], 0, 1)$ *be a Morse function, and suppose that* $0 < c < 1$ *where* c *is not a critical value of* f. *Then both* $f^{-1}[0,c]$ *and* $f^{-1}[c,1]$ *are smooth manifolds with boundary.*

Hence the cobordism $(W; V_0, V_1;$ identity, identity$)$ from V_0 to V_1 can be expressed as the composition of two cobordisms: one from V_0 to $f^{-1}(c)$ and one from $f^{-1}(c)$ to V_1. Together with 2.8 this proves:

Corollary 2.10. *Any cobordism can be expressed as a composition of cobordisms with Morse number* 1.

Proof of 2.9: This follows immediately from the implicit function theorem, for if $w \in f^{-1}(c)$, then, in some coordinate system $x_1, x_2, \ldots, x_n$ about w, f looks locally like the projection map $R^n \longrightarrow R$, $(x_1, \ldots, x_n) \longrightarrow x_n$.

Section 3. Elementary Cobordisms

Definition 3.1. Let f be a Morse function for the triad $(W^n; V, V')$. A vector field ξ on W^n is a gradient-like vector field for f if

1) $\xi(f) > 0$ throughout the complement of the set of critical points of f, and

2) given any critical point p of f there are coordinates $(\vec{x}, \vec{y}) = (x_1, \ldots, x_\lambda, x_{\lambda+1}, \ldots, x_n)$ in a neighborhood U of p so that $f = f(p) - |\vec{x}|^2 + |\vec{y}|^2$ and ξ has coordinates $(-x_1, \ldots, -x_\lambda, x_{\lambda+1}, \ldots, x_n)$ throughout U.

Lemma 3.2. *For every Morse function f on a triad $(W^n; V, V')$ there exists a gradient-like vector field ξ.*

Proof. For simplicity we assume f has only one critical point p, the proof in general being similar. By the Morse Lemma 2.2 we may choose coordinates $(\vec{x}, \vec{y}) = (x_1, \ldots, x_\lambda, x_{\lambda+1}, \ldots, x_n)$ in a neighborhood U_0 of p so that $f = f(p) - |\vec{x}|^2 + |\vec{y}|^2$ throughout U_0. Let U be a neighborhood of p such that $\bar{U} \subset U_0$.

Each point $p' \in W - U_0$ is not a critical point of f. It follows from the Implicit Function Theorem that there exist coordinates $x_1', \ldots, x_n'$ in a neighborhood U' of p' such that $f = \text{constant} + x_1'$ in U'.

Using this and the fact that $W - U_0$ is compact, find neighborhoods $U_1, \ldots, U_k$ such that

1) $W - U_0 \subset U_1 \cup \ldots \cup U_k$,

2) $U \cap U_i = \emptyset$, $i = 1,\ldots,k$, and

3) U_i has coordinates $x_1^i, \ldots, x_n^i$ and $f = \text{constant} + x_1^i$ on U_i, $i = 1,\ldots,k$.

On U_0 there is the vector field whose coordinates are $(-x_1, \ldots, -x_\lambda, x_{\lambda+1}, \ldots, x_n)$, and on U_i there is the vector field $\partial/\partial x_1^i$ with coordinates $(1,0,\ldots,0)$, $i = 1,\ldots,k$. Piece together these vector fields using a partition of unity subordinate to the cover $U_0, U_1, \ldots, U_k$, obtaining a vector field ξ on W . It is easy to check that ξ is the required gradient-like vector field for f .

Remark. From now on we shall identify the triad $(W; V_0, V_1)$ with the cobordism $(W; V_0, V_1; i_0, i_1)$ where $i_0: V_0 \longrightarrow V_0$ and $i_1: V_1 \longrightarrow V_1$ are the identity maps.

Definition 3.3. A triad $(W; V_0, V_1)$ is said to be a product cobordism if it is diffeomorphic to the triad $(V_0 \times [0,1]; V_0 \times 0, V_0 \times 1)$.

Theorem 3.4. *If the Morse number* μ *of the triad* $(W; V_0, V_1)$ *is zero, then* $(W; V_0, V_1)$ *is a product cobordism.*

Proof: Let $f: W \longrightarrow [0,1]$ be a Morse function with no critical points. By Lemma 3.2 there exists a gradient-like vector field ξ for f. Then $\xi(f): W \longrightarrow R$ is strictly positive. Multiplying ξ at each point by the positive real number $1/\xi(f)$, we may assume $\xi(f) = 1$ identically on W .

If p is any point in Bd W, then f expressed in some coordinate system $x_1, \ldots, x_n$, $x_n \geq 0$, about p extends to a smooth function g defined on an open subset U of R^n. Correspondingly, ξ expressed in this coordinate system also extends to U. The fundamental existence and uniqueness theorem for ordinary differential equations (see e.g. Lang [3, p.55]) thus applies locally to W.

Let $\varphi : [a, b] \longrightarrow W$ be any integral curve for the vector field ξ. Then

$$\frac{d}{dt}(f \circ \varphi) = \xi(f)$$

is identically equal to 1 ; hence

$$f(\varphi(t)) = t + \text{constant}.$$

Making the change of parameter, $\psi(s) = \varphi(s - \text{constant})$, we obtain an integral curve which satisfies

$$f(\psi(s)) = s .$$

Each integral curve can be extended uniquely over a maximal interval, which, since W is compact, must be $[0, 1]$. Thus, for each $y \in W$ there exists a unique maximal integral curve

$$\psi_y : [0, 1] \longrightarrow W$$

which passes through y, and satisfies $f(\psi_y(s)) = s$. Furthermore $\psi_y(s)$ is smooth as a function of both variables (cf. §5, pages 53 - 54).

The required diffeomorphism

$$h: V_0 \times [0,1] \longrightarrow W$$

is now given by the formula

$$h(y_0, s) = \psi_{y_0}(s) ,$$

with

$$h^{-1}(y) = (\psi_y(0), f(y)) .$$

Corollary 3.5. (Collar Neighborhood Theorem) Let W be a compact smooth manifold with boundary. There exists a neighborhood of Bd W (called a collar neighborhood) diffeomorphic to Bd $W \times [0,1)$.

Proof. By lemma 2.6, there exists a smooth function $f: W \longrightarrow R_+$ such that $f^{-1}(0) =$ Bd W and $df \neq 0$ on a neighborhood U of Bd W . Then f is a Morse function on $f^{-1}[0, \epsilon/2]$, where $\epsilon > 0$ is a lower bound for f on the compact set $W - U$. Thus Theorem 3.4 guarantees a diffeomorphism of $f^{-1}[0, \epsilon/2)$ with Bd $W \times [0,1)$.

A connected, closed submanifold $M^{n-1} \subset W^n -$ Bd W^n is said to be two-sided if some neighborhood of M^{n-1} on W^n is cut into two components when M^{n-1} is deleted.

Corollary 3.6. (The Bicollaring Theorem) Suppose that every component of a smooth submanifold M of W is compact and two-sided. Then there exists a "bicollar" neighborhood of M in W diffeomorphic to $M \times (-1,1)$ in such a way that M corresponds to $M \times 0$

Proof. Since the components of M may be covered by disjoint open sets in W , it suffices to consider the case where M has a single component.

Let U be an open neighborhood of M in W - Bd W such that $\bar{U}$ is compact and lies in a neighborhood of M which is cut into two components when M is deleted. Then U clearly splits up as a union of two submanifolds U_1, U_2 such that $U_1 \cap U_2 = M$ is the boundary of each. As in the proof of 2.6 one can use a coordinate cover and a partition of unity to construct a smooth map

$$\varphi: U \longrightarrow R$$

such that $d\varphi \neq 0$ on M , and $\varphi < 0$ on $\bar{U} - U_1$, $\varphi = 0$ on M , $\varphi > 0$ on $\bar{U} - U_2$. We can choose an open neighborhood V of M , with $\bar{V} \subset U$, on which φ has no critical points.

Let $2\epsilon'' > 0$ be the lub of φ on the compact set $\bar{U}_1 - V$.

Let $2\epsilon' < 0$ be the glb of φ on the compact set $\bar{U}_2 - V$.

Then $\varphi^{-1}[\epsilon',\epsilon'']$ is a compact n-dimensional sub-manifold of V with boundary $\varphi^{-1}(\epsilon') \cup \varphi^{-1}(\epsilon'')$, and φ is a Morse function on $\varphi^{-1}[\epsilon',\epsilon'']$. Applying Theorem 3.1 we find that $\varphi^{-1}(\epsilon',\epsilon'')$ is a "bicollar" neighborhood of M in V and so also in W .

Remark. The collaring and bicollaring theorems remain valid without the compactness conditions. (Munkres [5, p. 51]).

We now restate and prove a result of Section 1.

<u>Theorem 1.4. Let</u> $(W; V_0, V_1)$ <u>and</u> $(W'; V_1', V_2')$ <u>be two smooth manifold triads and</u> $h: V_1 \longrightarrow V_1'$ <u>a diffeomorphism. Then there exists a smoothness structure</u> $\mathcal{S}$ <u>for</u> $W \cup_h W'$ <u>compatible with the given structures on</u> W <u>and</u> W'. $\mathcal{S}$ <u>is unique up to a diffeomorphism leaving</u> V_0, $h(V_1) = V_1'$, <u>and</u> V_2' <u>fixed.</u>

<u>Proof</u>. <u>Existence</u>: By Corollary 3.5, there exist collar neighborhoods U_1, U_1' of V_1, V_1' in W, W' and diffeomorphisms $g_1: V_1 \times (0,1] \longrightarrow U_1$, $g_2: V_1' \times [1,2) \longrightarrow U_1'$, such that $g_1(x,1) = x$, $x \in V_1$, and $g_2(y,1) = y$, $y \in V_1'$. Let $j: W \longrightarrow W \cup_h W'$, $j': W' \longrightarrow W \cup_h W'$ be the inclusion maps in the definition of $W \cup_h W'$. Define a map $g: V_1 \times (0,2) \longrightarrow W \cup_h W'$ by

$$g(x,t) = j(g_1(x,t)) \qquad 0 < t \leq 1$$

$$g(x,t) = j'(g_2(h(x),t)) \quad 1 \leq t < 2 .$$

To define a smoothness structure on a manifold it suffices to define compatible smoothness structures on open sets covering the manifold. $W \cup_h W'$ is covered by $j(W - V_1)$, $j'(W' - V_1')$, and $g(V_1 \times (0,2))$, and the smoothness structures defined on these sets by j, j', and g respectively, are compatible. This completes the proof of existence.

<u>Uniqueness</u>: We show that any smoothness structure $\mathcal{S}$ on $W \cup_h W'$ compatible with the given structures on W and W' is isomorphic to a smoothness structure constructed by pasting together collar neighborhoods of V_1 and V_1' as above. The uniqueness up to diffeomorphism leaving V_0, $h(V_1) = V_1'$, and V_2' fixed then follows essentially from Theorem 6.3 of Munkres [5, p. 62]. By Corollary 3.6 there exists a

bicollar neighborhood U of $j(V_1) = j'(V_1')$ in $W \cup_h W'$ and a diffeomorphism $g: V_1 \times (-1,1) \longrightarrow U$ with respect to the smoothness structure $\mathcal{S}$, so that $g(x,0) = j(x)$, for $x \in V_1$. Then $j^{-1}(U \cap j(W))$ and $j'^{-1}(U \cap j'(W'))$ are collar neighborhoods of V_1 and V_1' in W and W'. This completes the proof of uniqueness.

Suppose now we are given triads $(W; V_0, V_1)$, $(W'; V_1', V_2')$ with Morse functions f, f' to $[0,1]$, $[1,2]$, respectively. Construct gradient-like vector fields ξ and ξ' on W and W', respectively, normalized so that $\xi(f) = 1$, $\xi'(f') = 1$ except in a small neighborhood of each critical point.

Lemma 3.7. *Given a diffeomorphism $h: V_1 \longrightarrow V_1'$ there is a unique smoothness structure on $W \cup_h W'$, compatible with the given structures on W, W', so that f and f' piece together to give a smooth function on $W \cup_h W'$ and ξ and ξ' piece together to give a smooth vector field.*

Proof. The proof is the same as that of Theorem 1.4 above, except that the smoothness structure on the bicollar neighborhood must be chosen by piecing together integral curves of ξ and ξ' in collar neighborhoods of V_1 and V_1'. This condition also proves uniqueness. (Notice that uniqueness here is much stronger than that in Theorem 1.4.)

This construction gives an immediate proof of the following result.

Corollary 3.8. $\mu(W \cup_h W'; V_0, V_2') \leq \mu(W; V_0, V_1) + \mu(W'; V_1', V_2')$ *where μ is the Morse number of the triad.*

Next we will study cobordisms with Morse number 1.

Let $(W; V, V')$ be a triad with Morse function $f: W \longrightarrow R$ and gradient-like vector field ξ for f. Suppose $p \in W$ is a critical point, and $V_0 = f^{-1}(c_0)$ and $V_1 = f^{-1}(c_1)$ are levels such that $c_0 < f(p) < c_1$ and that $c = f(p)$ is the only critical value in the interval $[c_0, c_1]$.

Let OD_r^p denote the open ball of radius r with center 0 in R^p, and set $OD_1^p = OD^p$.

Since ξ is a gradient-like vector field for f, there exists a neighborhood U of p in W, and a coordinate diffeomorphism $g: OD_{2\epsilon}^n \longrightarrow U$ so that $fg(\vec{x}, \vec{y}) = c - |\vec{x}|^2 + |\vec{y}|^2$ and so that ξ has coordinates $(-x_1, \ldots, -x_\lambda, x_{\lambda+1}, \ldots, x_n)$ throughout U, for some $-1 \leq \lambda \leq n$ and some $\epsilon > 0$. Here $\vec{x} = (x_1, \ldots, x_\lambda) \in R^\lambda$ and $\vec{y} = (x_{\lambda+1}, \ldots, x_n) \in R^{n-\lambda}$. Set $V_{-\epsilon} = f^{-1}(c - \epsilon^2)$ and $V_\epsilon = f^{-1}(c + \epsilon^2)$ We may assume $4\epsilon^2 < \min(|c - c_0|, |c - c_1|)$, so that $V_{-\epsilon}$ lies between V_0 and $f^{-1}(c)$ and V_ϵ lies between $f^{-1}(c)$ and V_1. The situation is represented schematically in Figure 3.1.

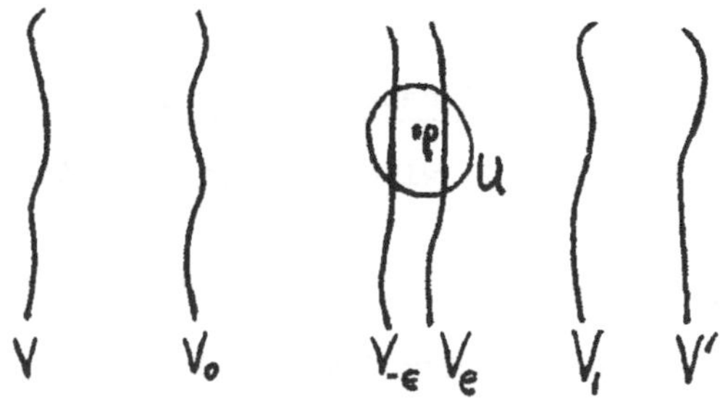

<u>Figure 3.1</u>

Let S^{p-1} denote the boundary of the closed unit disc D^p in R^p.

Definition 3.9. The characteristic embedding $\varphi_L\colon S^{\lambda-1} \times OD^{n-\lambda} \longrightarrow V_0$ is obtained as follows. First define an embedding $\varphi\colon S^{\lambda-1} \times OD^{n-\lambda} \longrightarrow V_{-\epsilon}$ by $\varphi(u, \theta v) = g(\epsilon u \cosh \theta, \epsilon v \sinh \theta)$ for $u \in S^{\lambda-1}$, $v \in S^{n-\lambda-1}$, and $0 \leq \theta < 1$. Starting at the point $\varphi(u, \theta v)$ in $V_{-\epsilon}$ the integral curve of ξ is a non-singular curve which leads from $\varphi(u, \theta v)$ back to some well-defined point $\varphi_L(u, \theta v)$ in V_0. Define the left-hand sphere S_L of p in V_0 to be the image $\varphi_L(S^{\lambda-1} \times 0)$. Notice that S_L is just the intersection of V_0 with all integral curves of ξ leading to the critical point p. The left hand disc D_L is a smoothly imbedded disc with boundary S_L, defined to be the union of the segments of these integral curves beginning in S_L and ending at p.

Similarly the characteristic embedding $\varphi_R\colon OD^{\lambda} \times S^{n-\lambda-1} \longrightarrow V_1$ is obtained by embedding $OD^{\lambda} \times S^{n-\lambda-1} \longrightarrow V_{\epsilon}$ by $(\theta u, v) \longrightarrow g(\epsilon u \sinh \theta, \epsilon v \cosh \theta)$ and then translating the image to V_1. The right-hand sphere S_R of p in V_1 is defined to be $\varphi_R(0 \times S^{n-\lambda-1})$. It is the boundary of the right-hand disk D_R, defined as the union of segments of integral curves of ξ beginning at p and ending in S_R.

Definition 3.10. An elementary cobordism is a triad $(W; V, V')$ possessing a Morse function f with exactly one critical point p.

Remark. It follows from 3.15 below that an elementary cobordism $(W; V, V')$ is not a product cobordism, and hence by 3.4 that the Morse number $\mu(W; V, V')$ equals one. Also 3.15 implies that the index of the elementary cobordism $(W; V, V')$, defined to be the index of p with respect to the Morse function f, is well-defined (i.e., independent of choice of f and p).

Figure 3.2 illustrates an elementary cobordism of dimension $n = 2$ and index $\lambda = 1$.

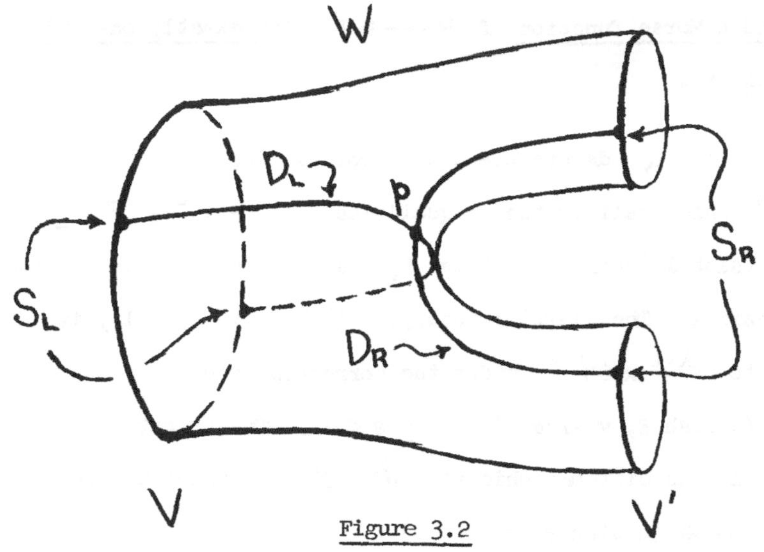

Figure 3.2

Definition 3.11. Given a manifold V of dimension $n-1$ and an embedding $\varphi: S^{\lambda-1} \times OD^{n-\lambda} \longrightarrow V$ let $\chi(V,\varphi)$ denote the quotient manifold obtained from the disjoint sum $(V - \varphi(S^{\lambda-1} \times 0)) + (OD^{\lambda} \times S^{n-\lambda-1})$ by identifying $\varphi(u, \theta v)$ with $(\theta u, v)$ for each $u \in S^{\lambda-1}$, $v \in S^{n-\lambda-1}$, $0 < \theta < 1$. If V' denotes any manifold diffeomorphic to $\chi(V, \varphi)$ then we will say that V' can be obtained from V by surgery of type $(\lambda, n-\lambda)$

Thus a surgery on an $(n-1)$-manifold has the effect of removing an embedded sphere of dimension $\lambda-1$ and replacing it by an embedded sphere of dimension $n-\lambda-1$. The next two results show that this corresponds to passing a critical point of index λ of a Morse function on an n-manifold.

<u>Theorem 3.12. If $V' = \chi(V,\varphi)$ can be obtained from V by surgery of type $(\lambda, n-\lambda)$, then there exists an elementary cobordism $(W; V, V')$ and a Morse function $f\colon W \longrightarrow R$ with exactly one critical point, of index λ.</u>

<u>Proof.</u> Let L_λ denote the set of points $(\vec{x}, \vec{y})$ in $R^\lambda \times R^{n-\lambda} = R^n$ which satisfy the inequalities $-1 \leq -|\vec{x}|^2 + |\vec{y}|^2 \leq 1$, and $|\vec{x}||\vec{y}| < (\sinh 1)(\cosh 1)$. Thus L_λ is a differentiable manifold with two boundaries. The "left" boundary, $-|\vec{x}|^2 + |\vec{y}|^2 = -1$, is diffeomorphic to $S^{\lambda-1} \times OD^{n-\lambda}$ under the correspondence $(u, \theta v) \longleftrightarrow (u \cosh \theta, v \sinh \theta)$, $0 \leq \theta < 1$. The "right" boundary, $-|\vec{x}|^2 + |\vec{y}|^2 = 1$, is diffeomorphic to $OD^\lambda \times S^{n-\lambda-1}$ under the correspondence $(\theta u, v) \longleftrightarrow (u \sinh \theta, v \cosh \theta)$.

Consider the orthogonal trajectories of the surfaces $-|\vec{x}|^2 + |\vec{y}|^2 = \text{constant}$. The trajectory which passes through the point $(\vec{x}, \vec{y})$ can be parametrized in the form $t \longrightarrow (t\vec{x}, t^{-1}\vec{y})$. If $\vec{x}$ or $\vec{y}$ is zero this trajectory is a straight line segment tending to the origin. For $\vec{x}$ and $\vec{y}$ different from zero it is a hyperbola which leads from some well-defined point $(u \cosh \theta, v \sinh \theta)$ on the left boundary of L_λ to the corresponding point $(u \sinh \theta, v \cosh \theta)$ on the right boundary.

Construct an n-manifold $W = \omega(V,\varphi)$ as follows. Start with the disjoint sum $(V - \varphi(S^{\lambda-1} \times 0)) \times D^1 + L_\lambda$. For each $u \in S^{\lambda-1}$, $v \in S^{n-\lambda-1}$, $0 < \theta < 1$, and $c \in D^1$ identify the point $(\varphi(u, \theta v), c)$ in the first summand with the unique point $(\vec{x}, \vec{y}) \in L_\lambda$ such that

(1) $-|\vec{x}|^2 + |\vec{y}|^2 = c$,

(2) $(\vec{x}, \vec{y})$ lies on the orthogonal trajectory which passes through the point $(u \cosh \theta, v \sinh \theta)$.

It is not difficult to see that this correspondence defines a diffeomorphism $\varphi(S^{\lambda-1} \times (OD^{n-\lambda} - 0)) \times D^1 \longleftrightarrow L_\lambda \cap (R^\lambda - 0) \times (R^{n-\lambda} - 0)$. It follows from this that $\omega(V,\varphi)$ is a well-defined smooth manifold.

This manifold $\omega(V,\varphi)$ has two boundaries, corresponding to the values $c = -|\vec{x}|^2 + |\vec{y}|^2 = -1$, and $+1$. The left boundary, $c = -1$, can be identified with V, letting $z \in V$ correspond to:

$$\begin{cases} (z, -1) \in (V - \varphi(S^{\lambda-1} \times 0)) \times D^1 & \text{for } z \notin \varphi(S^{\lambda-1} \times 0). \\ (u \cosh \theta, v \sinh \theta) \in L_\lambda & \text{for } z = \varphi(u, \theta v) . \end{cases}$$

The right boundary can be identified with $\chi(V,\varphi)$: letting $z \in V - \varphi(S^{\lambda-1} \times 0)$ correspond to $(z, +1)$; and letting $(\theta u, v) \in OD^\lambda \times S^{n-\lambda-1}$ correspond to $(u \sinh \theta, v \cosh \theta)$.

A function $f: \omega(V,\varphi) \longrightarrow R$ is defined by:

$$\begin{cases} f(z,c) = c & \text{for } (z,c) \in (V - \varphi(S^{\lambda-1} \times 0)) \times D^1. \\ f(\vec{x},\vec{y}) = -|\vec{x}|^2 + |\vec{y}|^2 & \text{for } (\vec{x},\vec{y}) \in L_\lambda \end{cases}$$

It is easy to check that f is a well-defined Morse function with one critical point, of index λ . This completes the proof of 3.12.

Theorem 3.13. *Let $(W; V, V')$ be an elementary cobordism with characteristic embedding $\varphi_L: S^{\lambda-1} \times OD^{n-\lambda} \longrightarrow V$. Then $(W; V, V')$ is diffeomorphic to the triad $(\omega(V, \varphi_L); V, \chi(V, \varphi_L))$.*

Proof. Using the notation of 3.9 with $V = V_0$ and $V' = V_1$, we know from 3.4 that $(f^{-1}([c_0, c-\epsilon^2]); V, V_{-\epsilon})$ and $(f^{-1}([c+\epsilon^2, c_1]); V_\epsilon, V')$ are product cobordisms. Thus $(W; V, V')$ is diffeomorphic to $(W_\epsilon; V_{-\epsilon}, V_\epsilon)$, where $W_\epsilon = f^{-1}([c-\epsilon^2, c+\epsilon^2])$. Since $(\omega(V,\varphi_L); V, \chi(V,\varphi_L))$ is clearly diffeomorphic to $(\omega(V_{-\epsilon},\varphi); V_{-\epsilon}, \chi(V_{-\epsilon},\varphi))$, it suffices to show $(W_\epsilon; V_{-\epsilon}, V_\epsilon)$ is diffeomorphic to $(\omega(V_{-\epsilon},\varphi); V_{-\epsilon}, \chi(V_{-\epsilon},\varphi))$.

Define a diffeomorphism $k: \omega(V_{-\epsilon}, \varphi) \longrightarrow W_\epsilon$ as follows. For each $(z,t) \in (V_{-\epsilon} - \varphi(S^{\lambda-1} \times 0)) \times D^1$ let $k(z,t)$ be the unique point of W_ϵ such that $k(z,t)$ lies on the integral curve which passes through the point z and such that $f(k(z,t)) = \epsilon^2 t + c$. For each $(\vec{x},\vec{y}) \in L_\lambda$ set $k(\vec{x},\vec{y}) = g(\epsilon\vec{x}, \epsilon\vec{y})$. It follows from the definitions of φ and of $\omega(V_{-\epsilon}, \varphi)$, and the fact that g sends orthogonal trajectories in L_λ to integral curves in W_ϵ, that we obtain a well-defined diffeomorphism from $\omega(V_{-\epsilon}, \varphi)$ to W_ϵ. This completes the proof of 3.13.

<u>Theorem 3.14</u>. <u>Let</u> $(W; V, V')$ <u>be an elementary cobordism possessing a Morse function with one critical point, of index</u> λ . <u>Let</u> D_L <u>be the left-hand disk associated to a fixed gradient-like vector field. Then</u> $V \cup D_L$ <u>is a deformation retract of</u> W.

<u>Corollary 3.15</u>. $H_*(W,V)$ <u>is isomorphic to the integers</u> Z <u>in dimension</u> λ <u>and is zero otherwise. A generator for</u> $H_\lambda(W,V)$ <u>is representel by</u> D_L.

<u>Proof of Corollary.</u>

We have
$$H_*(W,V) \cong H_*(V \cup D_L, V) \cong H_*(D_L, S_L) \cong \begin{cases} Z & \text{in dimension } \lambda \\ 0 & \text{otherwise} \end{cases}$$

where the second isomorphism is excision.

Proof of Theorem 3.14. By 3.13 we may assume that for the characteristic embedding $\varphi_L : S^{\lambda-1} \times OD^{n-\lambda} \longrightarrow V$ we have

$$W = \omega(V, \varphi_L) = (V - \varphi_L(S^{\lambda-1} \times 0)) \times D^1 + L_\lambda$$

modulo identifications, where now D_L is the disk

$$\{(\vec{x}, \vec{y}) \in L_\lambda \mid |\vec{y}| = 0\} .$$

Let

$$C = \{(\vec{x}, \vec{y}) \in L_\lambda \mid |\vec{y}| \leq \frac{1}{10}\}$$

be the $\frac{1}{10}$ cylindrical neighborhood of D_L.

We define deformation retractions r_t from W to $V \cup C$ and r_t' from $V \cup C$ to $V \cup D_L$. (Here $t \in [0, 1]$.) Composing these gives the desired retraction.

1st Retraction: Outside L_λ follow trajectories back to V. In L_λ follow them as far as C or V. Precisely:

For each $(v, c) \in (V - \varphi_L(S^{\lambda-1} \times OD^{n-\lambda})) \times D^1$ define $r_t(v, c) = (v, c - t(c+1))$.

For each $(\vec{x}, \vec{y}) \in L_\lambda$ define

$$r_t(\vec{x}, \vec{y}) = \begin{cases} (\vec{x}, \vec{y}) & \text{for } |\vec{y}| \leq \frac{1}{10} \\ (\frac{\vec{x}}{\rho}, \rho\vec{y}) & \text{for } |\vec{y}| \geq \frac{1}{10} \end{cases}$$

where $\rho = \rho(\vec{x}, \vec{y}, t)$ is the maximum of $1/(10|\vec{y}|)$ and the positive real solution for ρ of the equation

$$-\frac{|\vec{x}|^2}{\rho^2} + \rho^2|\vec{y}|^2 = [-|\vec{x}|^2 + |\vec{y}|^2](1 - t) - t .$$

Since for $|\vec{y}| \geq \frac{1}{10}$ the equation has a unique solution > 0 which varies continuously, it follows easily that r_t is a well-defined retraction from W to $V \cup C$.

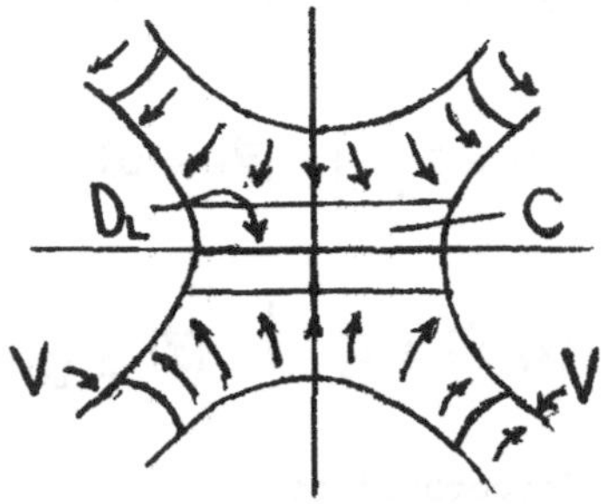

Figure 3.3. The retraction from W to $V \cup C$.

2nd Retraction. Outside of C define r_t' to be the identity. (Case 1).

In C move along straight lines vertically to $V \cup D_L$, moving more slowly near $V \cap C$. Precisely:

For each $(\vec{x}, \vec{y}) \in C$ define

$$r_t'(\vec{x}, \vec{y}) = \begin{cases} (\vec{x}, (1-t)\vec{y}) & \text{for } |\vec{x}|^2 \leq 1 \quad \text{(Case 2)} \\ (\vec{x}, \alpha\vec{y}) & \text{for } 1 \leq |\vec{x}|^2 \leq 1 + \frac{1}{100} \quad \text{(Case 3)} \end{cases}$$

where $\alpha = \alpha(\vec{x}, \vec{y}, t) = (1-t) + t((|\vec{x}|^2 - 1)/|\vec{y}|^2)^{1/2}$. One verifies that r_t' remains continuous as $|\vec{x}|^2 \to 1$, $|\vec{y}|^2 \to 0$. Note that the two definitions of r_t' agree for $|\vec{x}|^2 = 1$. This completes the proof of Theorem 3.

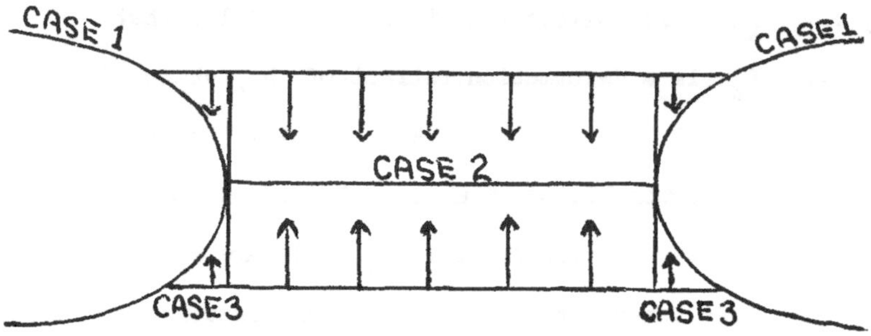

Figure 3.4. The retraction from $V \cup C$ to $V \cup D_L$.

Remark. We now indicate briefly how most of the above results can be generalized to the case of more than one critical point.

Suppose $(W; V, V')$ is a triad and $f: W \longrightarrow R$ a Morse function with critical points $p_1, \ldots, p_k$, all on the same level, of indices $\lambda_1, \ldots, \lambda_k$. Choosing a gradient-like vector field for f , we obtain disjoint characteristic embeddings $\varphi_i: S^{\lambda_i - 1} \times OD^{n-\lambda_i} \longrightarrow V$, $i = 1, \ldots, k$. Construct a smooth manifold $\omega(V; \varphi_1, \ldots, \varphi_k)$ as follows. Start with the disjoint sum $(V - \bigcup_{i=1}^{k} \varphi_i(S^{\lambda_i - 1} \times 0)) \times D^1 + L_{\lambda_1} + \ldots + L_{\lambda_k}$. For each $u \in S^{\lambda_i - 1}$, $v \in S^{n-\lambda_i - 1}$, $0 < \theta < 1$, and $c \in D^1$ identify the point $(\varphi_i(u, \theta v), c)$ in the first summand with the unique point $(\vec{x}, \vec{y}) \in L_{\lambda_i}$ such that

(1) $-|\vec{x}|^2 + |\vec{y}|^2 = c$, and

(2) $(\vec{x}, \vec{y})$ lies on the orthogonal trajectory which passes through the point $(u \cosh \theta, u \sinh \theta)$.

As in Theorem 3.13 one proves that W is diffeomorphic to $\omega(V; \varphi_1, \ldots, \varphi_k)$. It follows from this, as in 3.14, that $V \cup D_1 \cup \ldots \cup D_k$ is a deformation retract of W , where D_i denotes the left hand disk of p_i, $i = 1,\ldots,k$. Finally, if $\lambda_1 = \lambda_2 = \ldots = \lambda_k = \lambda$ then $H_*(W,V)$ is isomorphic to $Z \oplus \ldots \oplus Z$ (k summands) in dimension λ and is zero otherwise. Generators for $H_\lambda(W,V)$ are represented by $D_1, \ldots, D_k$. These generators of $H_\lambda(W, V)$ are actually completely determined by the given Morse function without reference to the given gradient-like vector field — see [4, p. 20].

Section 4. Rearrangement of Cobordisms

From now on we shall use c to denote a cobordism, rather than an equivalence class of cobordisms as in Section 1. If a composition cc' of two elementary cobordisms is equivalent to a composition dd' of two elementary cobordisms such that

$$\text{index}(c) = \text{index}(d')$$

and $$\text{index}(c') = \text{index}(d)$$

then we say that the composition cc' can be rearranged. When is this possible?

Recall that on the triad $(W; V_0, V_1)$ for cc' there exists a Morse function $f: W \longrightarrow [0,1]$ with two critical points p and p', $\text{index}(p) = \text{index}(c)$, $\text{index}(p') = \text{index}(c')$, such that $f(p) < \frac{1}{2} < f(p')$ Given a gradient-like vector field ξ for f, the trajectories from p meet $V = f^{-1}(\frac{1}{2})$ in an imbedded sphere S_R, called the right-hand sphere of p, and the trajectories going to p' meet V in an imbedded sphere S_L', called the left-hand sphere for p'. We state a theorem which guarantees that cc' can be rearranged if $S_R \cap S_L' = \emptyset$.

Theorem 4.1. Preliminary Rearrangement Theorem. Let $(W; V_0, V_1)$ be a triad with Morse function f having two critical points p, p'. Suppose that for some choice of gradient-like vector field ξ, the compact set K_p of points on trajectories going to or from p is disjoint from the compact set $K_{p'}$ of points on trajectories going to or from p'. If $f(W) = [0,1]$ and $a, a' \in (0,1)$, then there exists a new Morse function g such that

(a) ξ is a gradient-like vector field for g ,

(b) the critical points of g are still p, p' , and $g(p) = a$, $g(p') = a'$,

(c) g agrees with f near $V_0 \cup V_1$ and equals f plus a constant in some neighborhood of p and in some neighborhood of p' .

(See Figure 4.1)

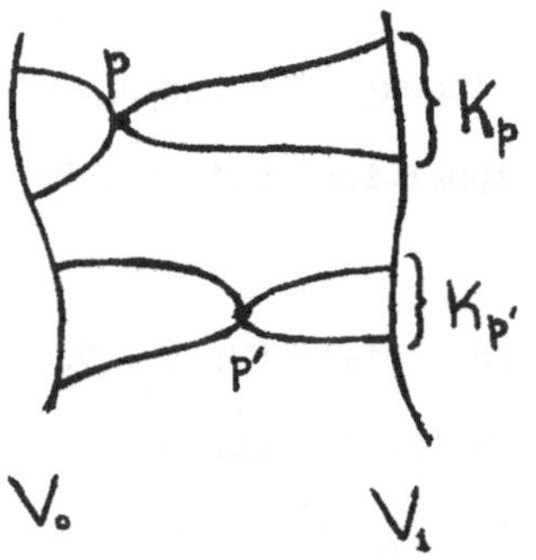

Figure 4.1

Proof: Clearly trajectories through points outside $K = K_p \cup K_{p'}$ all go from V_0 to V_1. The function $\pi\colon W - K \to V_0$ that assigns to each point q in $W - K$ the unique intersection of its trajectory with V_0 is smooth (cf. 3.4) and when q lies near K , then $\pi(q)$ lies near K in V_0. It follows that if $\mu\colon V_0 \to [0,1]$ is a smooth function zero near the left-hand sphere $K_p \cap V_0$, and one near the sphere $K_{p'} \cap V_0$, then μ extends uniquely to a smooth function $\bar{\mu}\colon W \to [0,1]$ that is constant on each trajectory, zero near K_p and one near $K_{p'}$.

Define a new Morse function $g\colon W \to [0,1]$ by $g(q) = G(f(q),\bar{\mu}(q))$ where $G(x,y)$ is any smooth function $[0,1] \times [0,1] \to [0,1]$ with the properties: (see Figure 4.2)

(i) For all x and y, $\frac{\partial G}{\partial x}(x,y) > 0$ and $G(x,y)$ increases from 0 to 1 as x increases from 0 to 1.

(ii) $G(f(p),0) = a$ $\qquad G(f(p'),1) = a'$

(iii) $G(x,y) = x$ $\qquad$ for x near 0 or 1 and for all y,

$\frac{\partial G}{\partial x}(x,0) = 1$ $\qquad$ for x in a neighborhood of $f(p)$,

$\frac{\partial G}{\partial x}(x,1) = 1$ $\qquad$ for x in a neighborhood of $f(p')$.

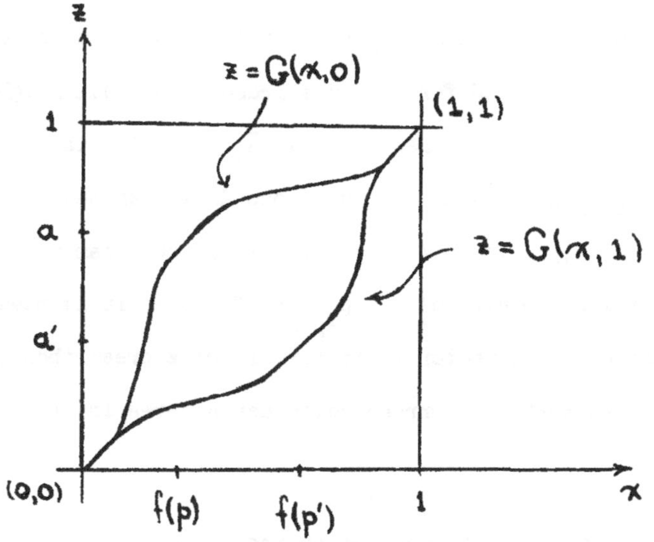

Figure 4.2

The reader can easily check that g has the desired properties (a), (b), and (c).

4.2. Extension: If more generally the Morse function f of 4.1 is allowed two sets of critical points $p = \{p_1,\ldots,p_n\}$, $p' = \{p'_1,\ldots,p'_\ell\}$ with all points of p at a single level $f(p)$ and all points of p' at a single level $f(p')$, then the theorem remains valid. In fact the proof may be repeated verbatim.

Still using the notation of page 37 let $\lambda = \text{index}(c)$, $\lambda' = \text{index}(c')$, and $n = \dim W$. If

$$\dim S_R + \dim S'_L < \dim V$$

i.e., $$(n - \lambda - 1) + (\lambda' - 1) < n-1$$

or $$\lambda \geq \lambda'$$

then, roughly stated, there is room enough to move S_R out of the way of S'_L .

<u>Theorem</u> 4.4. If $\lambda \geq \lambda'$, then it is possible to alter the gradient-like vector field for f on a prescribed small neighborhood of V so that the corresponding new spheres $\bar{S}_R$ and $\bar{S}'_L$ in V do not intersect. <u>More generally</u> if c is a cobordism with several index λ critical points $p_1,\ldots,p_k$ of f , and c' a cobordism with several index λ' critical points $p'_1,\ldots,p'_\ell$ of f , then it is possible to alter the gradient like vector field for f on a prescribed small neighborhood of V so that the corresponding new spheres in V are pairwise disjoint.

<u>Definition</u> 4.5. An open neighborhood U of a submanifold $M^m \subset V^v$, which is diffeomorphic to $M^m \times R^{v-m}$ in such a way that M^m corresponds to $M^m \times 0$, is called a <u>product neighborhood</u> of M^m in V^v .

<u>Lemma</u> 4.6. Suppose M and N are two submanifolds of dimension m and n in a manifold V of dimension v . If M has a product neighborhood in V , and $m+n < v$, then there exists a diffeomorphism h of V onto itself smoothly isotopic to the identity, such that $h(M)$ is disjoint from N .

<u>Remark</u>: The assumption that M has a product neighborhood is not necessary, but it simplifies the proof.

<u>Proof of 4.6</u>: Let $k: M \times R^{v-m} \longrightarrow U \subset V$ be a diffeomorphism onto a product neighborhood U of M in V such that $k(M \times \vec{0}) = M$. Let $N_0 = U \cap N$ and consider the composed map $g = \pi \circ k^{-1} | N_0$ where $\pi: M \times R^{v-m} \longrightarrow R^{v-m}$ is the natural projection.

The manifold $k(M \times \vec{x}) \subset V$ will intersect N if and only if $\vec{x} \in g(N_0)$. If N_0 is not empty, $\dim N_0 = n < v-m$; consequently the theorem of Sard (see de Rham [1,p.10]) shows that $g(N_0)$ has measure zero in R^{v-m}. Thus we may choose a point $\vec{u} \in R^{v-m} - g(N_0)$.

We will construct a diffeomorphism of V onto itself that carries M to $k(M \times \vec{u})$ and is isotopic to the identity. One can easily construct a smooth vector field $\zeta(\vec{x})$ on R^{v-m} such that $\zeta(\vec{x}) = \vec{u}$ for $|\vec{x}| \leq |\vec{u}|$ and $\zeta(\vec{x}) = 0$ for $|\vec{x}| \geq 2|\vec{u}|$. Since ζ has compact support, and R^{v-m} has no boundary, the integral curves $\psi(t,\vec{x})$ are defined for all real values of t. (Compare Milnor [4,p.10].) Then $\psi(0,\vec{x})$ is the identity on R^{v-m}, $\psi(1,\vec{x})$ is a diffeomorphism carrying 0 to $\vec{u}$, and $\psi(t,\vec{x})$, $0 \leq t \leq 1$, gives a smooth isotopy from $\psi(0,\vec{x})$ to $\psi(1,\vec{x})$.

Since this isotopy leaves all points fixed outside a bounded set in R^{v-m} we can use it to define an isotopy

$$h_t: V \longrightarrow V$$

by setting

$$h_t(w) = \begin{cases} k(q,\psi(t,\vec{x})) & \text{if } w = k(q,\vec{x}) \in U \\ w & \text{if } w \in V - U . \end{cases}$$

Then $h = h_1$ is the desired diffeomorphism $V \longrightarrow V$.

<u>Proof of Theorem 4.4</u>: To simplify notation we prove only the first statement of 4.4. The general statement is proved similarly.

Since the sphere S_R has a product neighborhood in V (cf. 3.9), Lemma 4.6 provides a diffeomorphism $h: V \longrightarrow V$ smoothly isotopic to the identity, for which $h(S_R) \cap S_L = \emptyset$. The isotopy is used as follows to alter ξ.

Let $a < \frac{1}{2}$ be so large that $f^{-1}[a,\frac{1}{2}]$ lies in the prescribed neighborhood of V. The integral curves of $\hat{\xi} = \xi/\xi(f)$ determine a diffeomorphism

$$\varphi: [a,\tfrac{1}{2}] \times V \longrightarrow f^{-1}[a,\tfrac{1}{2}]$$

such that $f(\varphi(t,q)) = t$, and $\varphi(\frac{1}{2},q) = q \in V$. Define a diffeomorphism H of $[a,\frac{1}{2}] \times V$ onto itself by setting $H(t,q) = (t,h_t(q))$, where $h_t(q)$ is a smooth isotopy $[a,\frac{1}{2}] \times V \longrightarrow V$ from the identity to h adjusted so that h_t is the identity for t near a and $h_t = h$ for t near $\frac{1}{2}$. Then one readily checks that

$$\xi' = (\varphi \circ H \circ \varphi^{-1})_* \hat{\xi}$$

is a smooth vector field defined on $f^{-1}[a,\frac{1}{2}]$ which coincides with $\hat{\xi}$ near $f^{-1}(a)$ and $f^{-1}(\frac{1}{2}) = V$, and satisfies $\xi'(f) = 1$ identically. Thus the vector field $\tilde{\xi}$ on W which coincides with $\xi(f)\xi'$ on $f^{-1}[a,\frac{1}{2}]$ and with ξ elsewhere is a new smooth gradient-like vector field for f.

<u>Figure 4.4.</u>

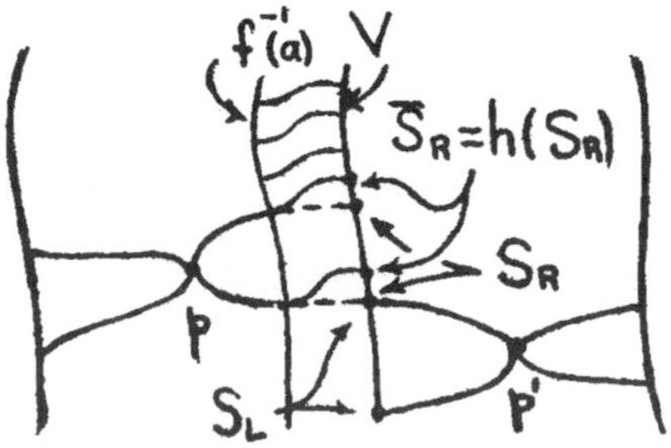

Now for each fixed $q \in V$, $\varphi(t, h_t(q))$ describes an integral curve of $\bar{\xi}$ from $\varphi(a,q)$ in $f^{-1}(a)$ to $\varphi(\frac{1}{2}, h(q)) = h(q)$ in $f^{-1}(\frac{1}{2}) = V$ It follows that the right-hand sphere $\varphi(a \times S_R)$ of p in $f^{-1}(a)$ is carried to $h(S_R)$ in V. Thus $h(S_R)$ is the new right-hand sphere $\bar{S}_R$ of p. Clearly $\bar{S}_L = S_L$. So $\bar{S}_R \cap \bar{S}_L = h(S_R) \cap S_L = \emptyset$ as required. This completes the proof of Theorem 4.4.

In the argument above we have proved the following lemma which is frequently needed in later sections.

Lemma 4.7. Given are a triad $(W; V_0, V_1)$ with Morse function f and gradient-like vector field ξ, a non-critical level $V = f^{-1}(b)$ and a diffeomorphism $h: V \longrightarrow V$ that is isotopic to the identity. If $f^{-1}[a,b]$, $a < b$, contains no critical points, then it is possible to construct a new gradient-like vector field $\bar{\xi}$ for f such that

(a) $\bar{\xi}$ coincides with ξ outside $f^{-1}(a,b)$

(b) $\bar{\varphi} = h \circ \varphi$, where φ and $\bar{\varphi}$ are the diffeomorphisms $f^{-1}(a) \longrightarrow V$ determined by following the trajectories of ξ and $\bar{\xi}$, respectively.

Replacing f by $-f$ one deduces a similar proposition in which ξ is altered on $f^{-1}(b,c)$, $b < c$, a neighborhood to the right rather than to the left of V.

Recall that any cobordism c may be expressed as a composition of a finite number of elementary cobordisms (Corollary 2.11). Applying the Preliminary Rearrangement Theorem 4.1, 4.2 in combination with Theorem 4.4 we obtain

Theorem 4.8. Final Rearrangement Theorem. Any cobordism c may be expressed as a composition

$$c = c_0 c_1 \cdots c_n , \quad n = \dim c ,$$

where each cobordism c_k admits a Morse function with just one critical level and with all critical points of index k.

Alternate version of 4.8.
Without using the notion of cobordism, we have the following proposition about Morse functions: Given any Morse function on a triad $(W; V_0, V_1)$, there exists a new Morse function f, which has the same critical points each with the same index, and which has the properties:

(1) $f(V_0) = -\frac{1}{2}$, $f(V_1) = n + \frac{1}{2}$

(2) $f(p) = \text{index}(p)$, at each critical point p of f.

Definition 4.9. Such a Morse function will be called self-indexing (or nice).

Theorem 4.8 is due to Smale [8] and Wallace [9].

In view of the Final Rearrangement Theorem another question arises naturally. When is a composition cc' of an elementary cobordism of index λ with an elementary cobordism of index $\lambda + 1$ equivalent to a product cobordism? Figure 5.1 shows how this may occur in dimension 2.

Figure 5.1

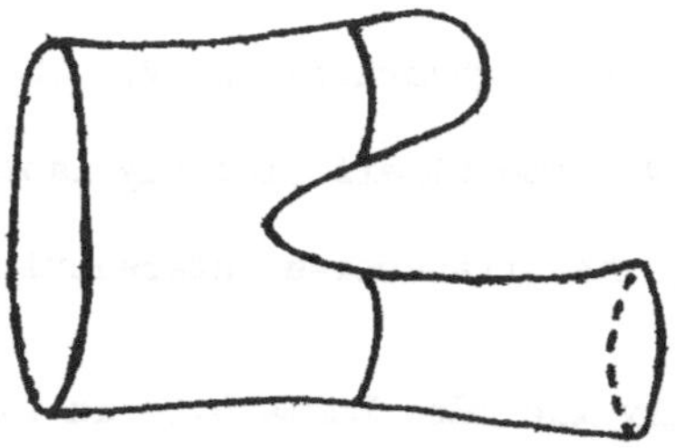

Let f be a Morse function on the triad $(W^n; V_o, V_1)$ for cc', having critical points p, p' of index $\lambda, \lambda + 1$ such that $f(p) < 1/2 < f(p')$. A gradient-like vector field ξ for f determines in $V = f^{-1}(1/2)$ a right-hand sphere S_R of p and a left-hand sphere S_L' of p'. Note that $\dim S_R + \dim S_L' = (n - \lambda - 1) + \lambda = n - 1 = \dim V$.

Definition 5.1

Two submanifolds $M^m, N^n \subset V^v$ are said to have transverse intersection (or to intersect transversely) if at each point $q \in M \cap N$ the tangent space to V at q is spanned by the vectors tangent to M and the vectors tangent to N. (If $m + n < v$ this is impossible, so transverse intersection simply means $M \cap N = \emptyset$.)

As a preliminary to the major Theorem 5.4 we prove:

Theorem 5.2

The gradient-like vector field ξ may be so chosen that S_R has transverse intersection with S'_L in V.

For the proof we use a lemma stated with the notation of Definition 5.1:

Lemma 5.3 If M has a product neighborhood in V, then there is a diffeomorphism h of V onto itself smoothly isotopic to the identity such that $h(M)$ has transverse intersection with N.

Remark: This lemma apparently includes Lemma 4.6; in fact the proof is virtually the same. The product neighborhood assumed for M is actually unnecessary.

Proof: As in Lemma 4.6 let $k : M \times R^{v-m} \longrightarrow U \subset V$ be a diffeomorphism onto a product neighborhood U of M in V such that $k(M \times \vec{o}) = M$. Let $N_o = U \cap N$, and consider the composed map $g = \pi \circ k^{-1}|N_o$ where $\pi : M \times R^{v-m} \longrightarrow R^{v-m}$ is the natural projection.

The manifold $k(M \times \vec{x})$ will fail to have transverse intersection with N if and only if $\vec{x} \varepsilon R^{v-m}$ is the image under g of some critical point $q \varepsilon N_o$ at which g fails to have maximal rank $v - m$. But according to the theorem of Sard (see Milnor [10, p. 10] and deRham [1, p.10]) the image $g(C)$ of the set $C \subset N_o$ of all critical points of g has measure zero in R^{v-m}. Hence we can choose a point $\vec{u} \varepsilon R^{v-m} - g(C)$, and, as in Lemma 4.6, construct an isotopy of the identity map of V to a diffeomorphism h of V onto itself that carries M to $k(M \times \vec{u})$. Since $k(M \times \vec{u})$ meets N transversely, the proof is complete.

Proof of Theorem 5.2:

The above lemma provides a diffeomorphism $h : V \longrightarrow V$ smoothly isotopic to the identity, such that $h(S_R)$ intersects S_L' transversely. Using Lemma 4.7 we can alter the gradient field ξ so that the new right-hand sphere is $h(S_R)$, and the left-hand sphere is unchanged. This completes the proof.

In the remainder of §5 it will be assumed that S_R has transverse intersection with S_L'. Since $\dim S_R + \dim S_L' = \dim V$, the intersection will consist of a finite number of isolated points. For if q_0 is in $S_R \cap S_L'$ there exist local coordinate functions $x^1(q), \ldots, x^{n-1}(q)$ on a neighborhood U of q_0 in V such that $x^i(q_0) = 0$, $i = 1, \ldots, n-1$, and $U \cap S_R$ is the locus $x^1(q) = \ldots = x^\lambda(q) = 0$ while $U \cap S_L'$ is the locus $x^{\lambda+1}(q) = \ldots = x^{n-1}(q) = 0$. Clearly the only point in $S_R \cap S_L' \cap U$ is q_0. As a consequence there are just a finite number of trajectories going from p to p', one through each point of $S_R \cap S_L'$.

Still using the notations introduced on page 45 we now state the major theorem of this section.

<u>Theorem 5.4</u> <u>First Cancellation Theorem</u>

If the intersection of S_R with S_L' is transverse and consists of a single point, then the cobordism is a product cobordism. In fact it is possible to alter the gradient-like vector field ξ on an arbitrarily small neighborhood of the single trajectory T from p to p' producing a nowhere zero vector field ξ' whose trajectories all proceed from V_0 to V_1. Further ξ' is a gradient-like vector field for a Morse function f' without critical points that agrees with f near $V_0 \cup V_1$.
(See Figure 5.2 below.)

<u>Remark</u>: The proof, due to M. Morse [11][32], is quite formidable. Not including the technical theorem 5.6 it occupies the following 10 pages.

Figure 5.2

Before Alteration

After

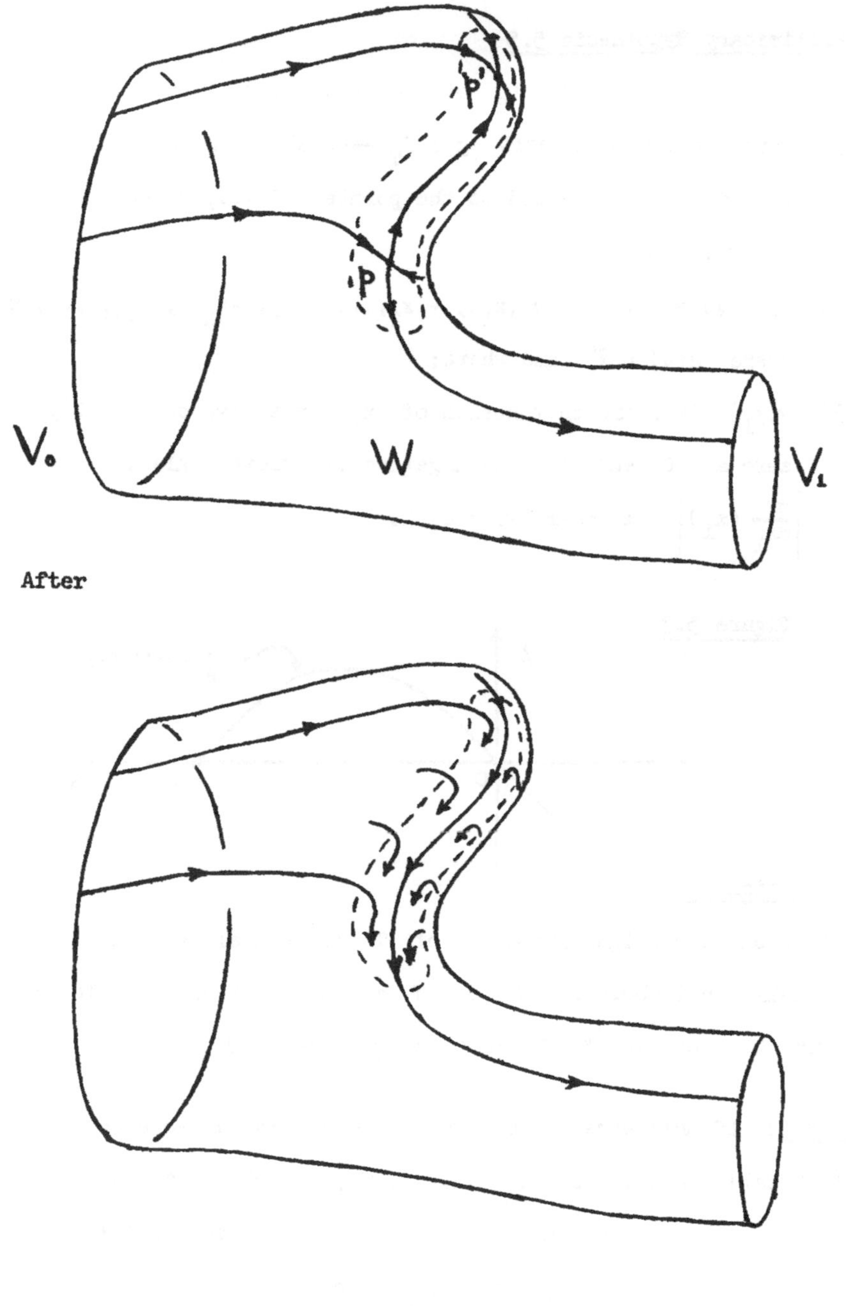

First we prove the theorem making an assumption about the behavior of ξ near T.

Preliminary Hypothesis 5.5

There is a neighborhood U_T of the trajectory T from p to p', and a coordinate chart $g : U_T \longrightarrow R^n$ such that:

1) p and p' correspond to the points $(0,\ldots,0)$ and $(1, 0,\ldots,0)$.
2) $g_* \xi(q) = \vec{\eta}(\vec{x}) = (v(x_1), -x_2,\ldots, -x_\lambda, -x_{\lambda+1}, x_{\lambda+2},\ldots, x_n)$ where $g(q) = \vec{x}$, and where:
3) $v(x_1)$ is a smooth function of x_1, positive on $(0, 1)$, zero at 0 and 1, and negative elsewhere. Also, $\left|\frac{\partial v}{\partial x_1}(x_1)\right| = 1$ near $x_1 = 0, 1$.

Figure 5.3

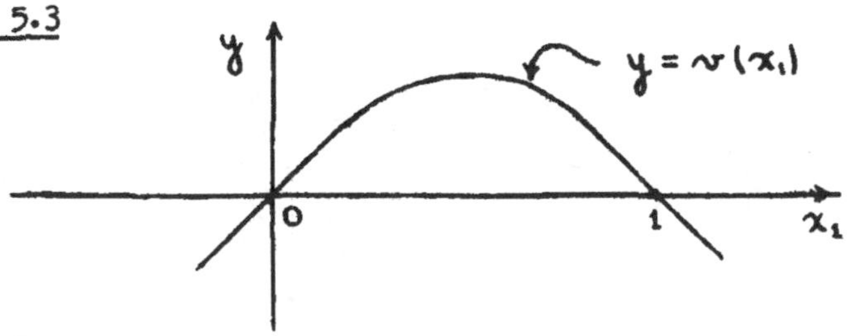

Assertion 1)

Given an open neighborhood U of T one can always find in U a smaller neighborhood U' of T so that no trajectory leads from U' outside of U and back again into U'.

Proof: If this were not so, there would exist a sequence of (partial) trajectories $T_1, T_2,\ldots, T_k,\ldots$ where T_k goes from a point r_k, through a point s_k outside U to a point t_k, and both sequences $\{r_k\}$ and $\{t_k\}$ approach T. Since W - U is compact we may assume that s_k converges to $s \in W - U$. The integral curve $\psi(t, s)$ through s must come from V_0 or go to V_1 or do both, else it would be a second trajectory joining p to p'. Suppose for definiteness that it comes from V_0. Then using the continuous dependence of $\psi(t, s')$ on the initial value s', we find that the trajectories through all points near s originate at V_0. The partial trajectory $T_{s'}$ from V_0 to any point s' near s is compact; hence the least distance $d(s')$ from T to $T_{s'}$ (in any metric) depends continuously on s' and will be bounded away from 0 for all s' in some neighborhood of s. Since $r_k \in T_{s_k}$ the points r_k cannot approach T as $k \longrightarrow \infty$, a contradiction.

Let U be any open neighborhood of T such that $\bar{U} \subset U_T$ and let U' be a 'safe' neighborhood, $T \subset U' \subset U$, provided by Assertion 1).

<u>Assertion 2)</u>

It is possible to alter ξ on a compact subset of U' producing a nowhere zero vector field ξ', such that every integral curve of ξ' through a point in U was outside U at some time $t' < 0$ and will again be outside U at some time $t'' > 0$.

<u>Proof</u>:

Replace $\vec{\eta}(\vec{x}) = (v(x_1), -x_2, \ldots, x_n)$ by a smooth vector field $\vec{\eta}'(\vec{x}) = (v'(x_1, \rho), -x_2, \ldots, x_n)$ where $\rho = [x_2^2 + \ldots + x_n^2]^{1/2}$

and

(i) $v'(x_1, \rho(\vec{x})) = v(x_1)$ outside a compact neighborhood of $g(T)$ in $g(U')$.

(ii) $v'(x_1, 0)$ is everywhere negative.

(see Figure 5.4)

Figure 5.4

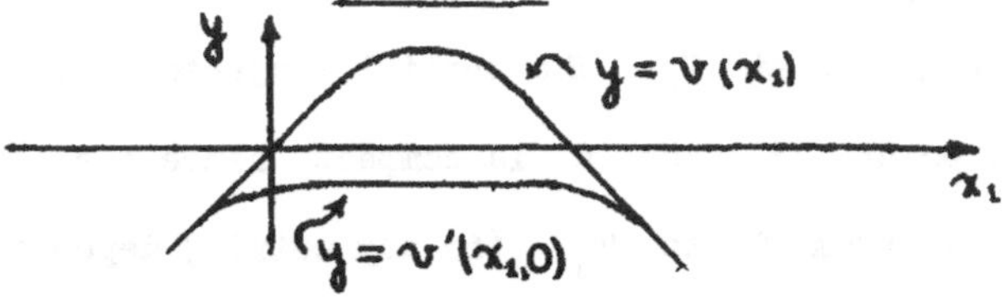

This determines a nowhere zero vector field ξ' on W. In our local coordinates, the differential equations satisfied by the integral curves of ξ' on U_T are

$$\frac{dx_1}{dt} = v'(x_1, \rho) , \frac{dx_2}{dt} = -x_2, \ldots, \frac{dx_{\lambda+1}}{dt} = -x_{\lambda+1} ,$$

$$\frac{dx_{\lambda+2}}{dt} = x_{\lambda+2}, \ldots, \frac{dx_n}{dt} = x_n$$

Consider the integral curve $\vec{x}(t) = (x_1(t), \ldots, x_n(t))$ with initial value $(x_1^o, \ldots, x_n^o)$, as t increases.

(a) If one of $x_{\lambda+2}^o, \ldots, x_n^o$ is nonzero, say $x_n^o \neq 0$, then $|x_n(t)| = |x_n^o e^t|$ increases exponentially and $\vec{x}(t)$ eventually leaves $g(U)$ ($g(\overline{U})$ is compact, therefore bounded).

(b) If $x_{\lambda+2}^o = \ldots = x_n^o = 0$, then $\rho(\vec{x}(t)) = [(x_2^o)^2 + \ldots + (x_{\lambda+1}^o)^2]^{1/2} e^{-t}$ decreases exponentially. Suppose $\vec{x}(t)$ remains in $g(U)$. Since $v'(x_1, \rho(\vec{x}))$ is negative on the x_1-axis, there exists $\delta > 0$ so small that $v'(x_1, \rho(\vec{x}))$ is negative on the compact set $K_\delta = \{\vec{x} \in g(\overline{U}) | \rho(\vec{x}) \leq \delta\}$.

Then $v'(x_1, \rho(\vec{x}))$ has a negative upperbound $-\alpha < 0$ on K_δ.

Eventually $\rho(\vec{x}(t)) \leq \delta$, and thereafter

$$\frac{dx_1(t)}{dt} \leq -\alpha .$$

Thus $\vec{x}(t)$ must eventually leave the bounded set $g(U)$ after all.

A similar argument will show that $\vec{x}(t)$ goes outside $g(U)$ as t decreases.

Assertion 3)

Every trajectory of the vector field ξ' goes from V_0 to V_1.

Proof:

If an integral curve of ξ' is ever in U' it eventually gets outside U, by Assertion 2). Leaving U' it follow trajectories of ξ; so once out of U it will remain out of U' permanently by Assertion 1). Consequently it must follow a trajectory of ξ to V_1. A parallel argument shows that it comes from V_0. On the other hand if an integral curve of ξ' is never in U' it is an integral curve of ξ that goes from V_0 to V_1.

Assertion 4) In a natural way, ξ' determines a diffeomorphism $\phi : ([0, 1] \times V_0 ; 0 \times V_0, 1 \times V_0) \longrightarrow (W; V_0, V_1)$

Proof: Let $\psi(t, q)$ be the family of integral curves for ξ'. Since ξ' is nowhere tangent to BdW, an application of the implicit function theorem shows that the function $\tau_1(q)$ [respectively $\tau_0(q)$] that assigns to each point $q \in W$ the time at which $\psi(t, q)$ reaches V_1 [respectively minus the time when it reaches V_0] depends smoothly on q. Then the projection $\pi : W \longrightarrow V_0$ given by $\pi(q) = \psi(-\tau_0(q), q)$ is also smooth. Clearly the smooth vector field $\tau_1(\pi(q))\, \xi'(q)$ has integral curves that go from V_0 to V_1 in unit time. To simplify notation assume that ξ' had this property from the outset. Then the required diffeomorphism ϕ maps

$$(t, q_0) \longrightarrow \psi(t, q_0)$$

and its inverse is the smooth map

$$q \longrightarrow (\tau_0(q), \pi(q)).$$

Assertion 5) The vector field ξ' is a gradient-like vector field for a Morse function g on W (with no critical points) that agrees with f on a neighborhood of $V_0 \cup V_1$.

Proof: In view of Assertion 4) it will suffice to exhibit a Morse function $g : [0, 1] \times V_0 \longrightarrow [0, 1]$ such that $\frac{\partial g}{\partial t} > 0$ and g agrees with $f_1 = f \circ \phi$ near $0 \times V_0 \cup 1 \times V_0$ (we may assume that $V_0 = f^{-1}(0)$ and $V_1 = f^{-1}(1)$). Clearly there exists $\delta > 0$ such that, for all $q \in V_0$, $\frac{\partial f_1}{\partial t}(t, q) > 0$ if $t < \delta$ or $t > 1 - \delta$. Let $\lambda : [0, 1] \longrightarrow [0, 1]$ be a smooth function zero for $t \in [\delta, 1 - \delta]$ and one for t near 0 and 1. Consider the function

$$g(u, q) = \int_0^u \{\lambda(t) \frac{\partial f_1}{\partial t}(t, q) + [1 - \lambda(t)]k(q)\}dt$$

where $k(q) = \{1 - \int_0^1 \lambda(t) \frac{\partial f_1}{\partial t}(t, q)dt\} / \int_0^1 [1 - \lambda(t)]dt.$

Choosing δ sufficiently small we may assume that $k(q) > 0$ for all $q \in V_0$. Then g apparently has the required properties.

Granting the Preliminary Hypothesis, this completes the proof of the First Cancellation Theorem 5.5. To establish Theorem 5.5 in general it remains to prove:

Assertion 6) When S_R and S_L' have a single, transverse intersection it is always possible to choose a new gradient-like vector field ξ' so that the Preliminary Hypothesis 5.5 is satisfied.

Remark: The proof, which occupies the last 12 pages of this section, has two parts — the reduction of the problem to a technical lemma (Theorem 5.6), and the proof of the lemma.

Proof: Let $\vec{\eta}(\vec{x})$ be a vector field on R^n that is of the form described in the Preliminary Hypothesis, with singularities at the origin 0 and the unit point e of the x_1-axis. The function

$$F(\vec{x}) = f(p) + 2\int_0^{x_1} v(t)dt - x_1^2 - \dots - x_{\lambda+1}^2 + x_{\lambda+2}^2 + \dots + x_n^2$$

is a Morse function on R^n for which $\vec{\eta}(\vec{x})$ is a gradient-like vector field. By a suitable choice of the function $v(x_1)$ we may arrange that $F(e) = f(p')$, i.e. $2\int_0^1 v(t)dt = f(p') - f(p)$.

Recall that according to the definition 3.1 of a gradient-like vector field for f, there exists a co-ordinate system $(x_1, \dots, x_n)$ about each of the critical points p and p' in which f corresponds to a function $\pm x_1^2 \pm \dots \pm x_n^2$ of suitable index, and ξ has coordinates $(\pm x, \dots, \pm x_n)$ Then one readily checks that there exist levels b_1 and b_2,

$a_1 = f(p) < b_1 < b_2 < f(p') = a_2$, and diffeomorphisms g_1, g_2 of closed, disjoint neighborhoods L_1, L_2 of 0 and e onto neighborhoods of p and p' respectively such that:

(a) The diffeomorphisms carry $\vec{\eta}$ to ξ, F to f, and points on the segment oe to points on T.

(b) Let p_i denote $T \cap f^{-1}(b_i)$, i = 1, 2. The image of L_1 is a neighborhood in $f^{-1}[a_1, b_1]$ of the segment pp_1 of T, while the image of L_2 is a neighborhood in $f^{-1}[b_2, a_2]$ of the segment p_2p' of T (see Figure 5.5).

Figure 5.5

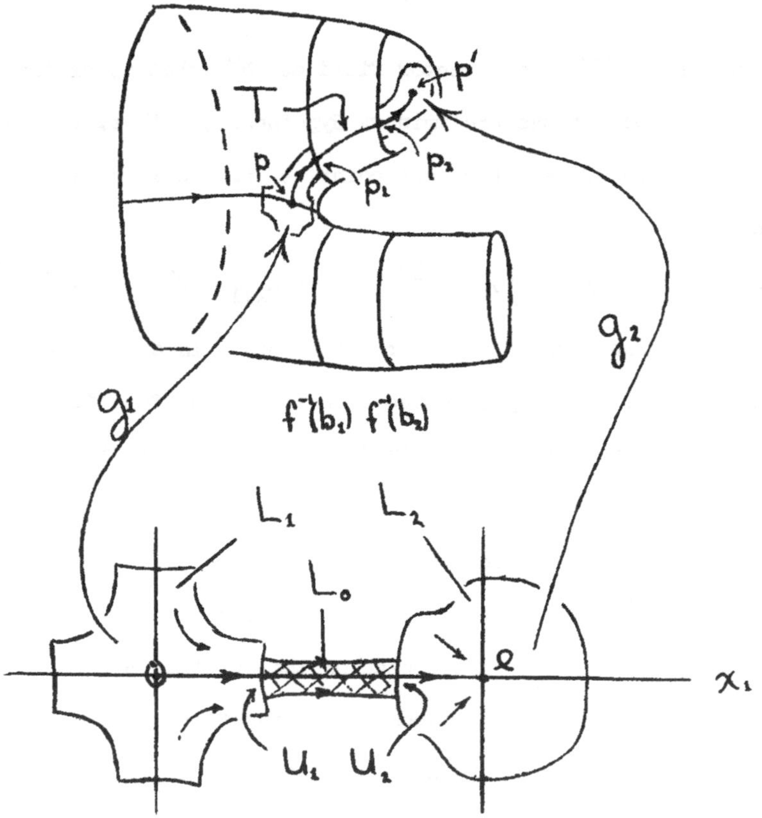

Observe that the trajectories of $\vec{\eta}(\vec{x})$ with initial points in a small neighborhood U_1 of $g_1^{-1}(p_1)$ in $g_1^{-1}f^{-1}(b_1)$ proceed to points in $g_2^{-1}f^{-1}(b_2)$ that form a diffeomorphic image U_2 of U_1 and in doing so sweep out a set L_o diffeomorphic to $U_1 \times [0, 1]$ such that $L_1 \cup L_o \cup L_2$ is a neighborhood of oe. There is a unique extension of g_1 to a smooth imbedding $\bar{g}_1$ of $L_1 \cup L_o$ into W determined by the condition that $\vec{\eta}$ trajectories go to ξ trajectories and F levels go to f levels.

Now let us suppose for the moment that the two imbeddings of U_2 into $f^{-1}(b_2)$ given by $\bar{g}_1$ and g_2 coincide at least on some small neighborhood of $g_2^{-1}(p_2)$ in U_2. Then $\bar{g}_1$ and g_2 together give a diffeomorphism $\bar{g}$ of a small neighborhood V of oe onto a neighborhood of T in W that preserves trajectories and levels. This implies that there is a smooth, positive, real-valued function k defined on $\bar{g}(V)$ such that for all points in $\bar{g}(V)$

$$\bar{g}_* \vec{\eta} = k \, \xi .$$

Choosing the neighborhood V of oe sufficiently small we may assume that the function k is defined, smooth and positive on all of W. Then $\xi' = k\xi$ is a gradient-like vector field satisfying the Preliminary Hypothesis 5.5. So when the above supposition holds the proof of Assertion 6) is complete.

In the general case, the vector field ξ determines a diffeomorphism $h : f^{-1}(b_1) \longrightarrow f^{-1}(b_2)$ and the vector field $\vec{\eta}$ determines a diffeomorphism $h' : U_1 \longrightarrow U_2$. Clearly the supposition made in the previous paragraph holds if and only if h coincides with $h_o = g_2 h' g_1^{-1}$ near p_1. Now, by Lemma 4.7, any diffeomorphism isotopic to h corresponds to a new gradient-like vector field that differs from ξ only on $f^{-1}(b_1, b_2)$. Thus Assertion 6) will be established if H can be deformed to a diffeomorphism $\bar{h}$ which coincides with h_o near p_1 and for which the new right-hand sphere $\bar{h}(S_R(b_1))$ in level b_2 still has the single transverse intersection p_2 with $S_L'(b_2)$. (The b_1 or b_2 here indicates the level in which the sphere lies.)

For convenience we will specify the required deformation of h by giving a suitable isotopy of $h_o^{-1}h$ that deforms $h_o^{-1}h$ on a very small neighborhood of p_1 to coincide with the identity map on a still smaller neighborhood of p_1. Observe that, after a preliminary alteration of g_2 if necessary, $h_o^{-1}h$ is orientation preserving at $p_1 = h_o^{-1}h(p_1)$ and both $h_o^{-1}h\ S_R(b_1)$ and $S_R(b_1)$ have the same intersection number (both $+1$ or both -1) with $S_L(b_1)$ at p_1. (For a definition of intersection number see §6.) Then the following local theorem provides the required isotopy.

Let $n = a + b$. A point $x \in R^n$ may be written $x = (u, v)$, $u \in R^a$, $v \in R^b$. We identify $u \in R^a$ with $(u, o) \in R^n$ and $v \in R^b$ with $(o, v) \in R^n$.

<u>Theorem 5.6</u>

Suppose that h is an orientation-preserving imbedding

of R^n into R^n such that

1) $h(0) = 0$ (where 0 denotes the origin in R^n)

2) $h(R^a)$ meets R^b only at the origin. The intersection is transverse and the intersection number is +1 (where we agree that R^a meets R^b with intersection number +1).

Then given any neighborhood N of the origin, there exists a smooth isotopy $h'_t : R^n \longrightarrow R^n$, $0 \leq t \leq 1$, with $h'_0 = h$ such that

(I) $h'_t(x) = h(x)$ for $x = 0$ and for $x \in R^n - N$, $0 \leq t \leq 1$.

(II) $h'_1(x) = x$ for x in some small neighborhood N_1 of 0.

(III) $h'_1(R^a) \cap R^b = 0$.

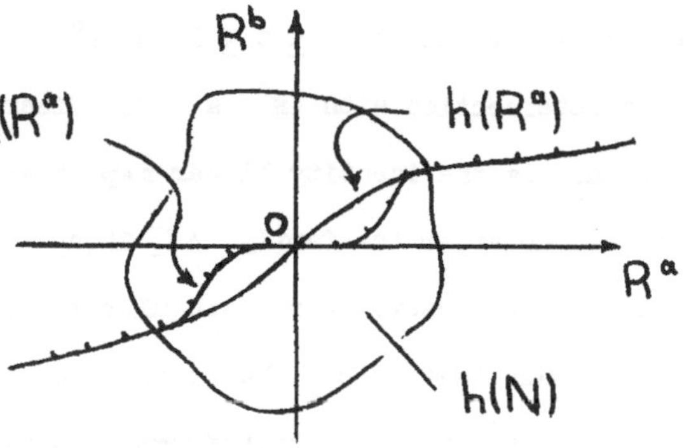

Figure 5.6

Lemma 5.7

Let $h : R^n \longrightarrow R^n$ be the map in the hypothesis of Theorem 5.6. There exists a smooth isotopy $h_t : R^n \longrightarrow R^n$, $0 \leq t \leq 1$, such that

(i) $h_0 = h$ and h_1 is the identity map of R^n.

(ii) for each $t \in [0, 1]$, $h_t(R^a) \cap R^b = 0$, and the intersection is transverse.

Proof of Lemma 5.7:

Since $h(0) = 0$, $h(x)$ may be expressed in the form $h(x) = x_1h^1(x) + \ldots + x_nh^n(x)$, $x = (x_1, \ldots, x_n)$, where $h^i(x)$ is a smooth vector function of x and (consequently) $h^i(0) = \frac{\partial h}{\partial x_i}(0)$, $i = 1, \ldots, n$. (see Milnor [4, p.6]). If we define h_t by

$$h_{1-t}(x) = \frac{1}{t}h(tx) = x_1h^1(tx) + \ldots + x_nh^n(tx),\ 0 \leq t \leq 1,$$

then $h_t(x)$ is clearly a smooth isotopy of h to the linear map

$$h_1(x) = x_1h^1(0) + \ldots + x_nh^n(0)$$

Since $h(R^a)$ and $h_t(R^a)$ have precisely the same orienting basis $h^1(0), \ldots, h^a(0)$ of tangent vectors at $0 \in R^n$, it follows that for all t, $0 \leq t \leq 1$, $h_t(R^a)$ has transverse positive intersection with R^b at 0. Clearly $h_t(R^a) \cap R^b = 0$. Thus if h_1 is the identity linear map we are through.

If not, consider the family $\Lambda \subset GL(n, R)$ consisting of all orientation-preserving non-singular linear transformations L of R^n such that $L(R^a)$ has transverse positive intersection with R^b, i.e. all transformations with matrices of the form $L = \left(\begin{array}{c|c} A & * \\ \hline * & * \end{array}\right)$

where A is an $a \times a$ matrix and

$$\det L > 0\ ,\ \det A > 0.$$

Assertion: For any $L \in \Lambda$ there is a smooth isotopy L_t, $0 \leq t \leq 1$, deforming L into the identity, such that $L_t \in \Lambda$ for all t; or, equivalently, there is a smooth path in Λ from L to the identity.

Proof: Addition of a scalar multiple of one of the first a rows [columns] to one of the last b rows [respectively, columns] clearly may be realized by a smooth deformation (= path) in $\mathcal{A}$. A finite number of such operations will reduce the matrix L to the form

$$L' = \left(\begin{array}{c|c} A & 0 \\ \hline 0 & B \end{array}\right)$$

where B is a $b \times b$ matrix and (necessarily) $\det B > 0$. As is well known a finite number of elementary operations on the matrix A, each realizable by a deformation in $GL(a, R)$, serve to reduce A to the identity matrix. A similar statement holds for B. Thus there are smooth deformations A_t, B_t, $0 \leq t \leq 1$, of A and B to identity matrices with $\det A_t > 0$ and $\det B_t > 0$. They provide a deformation in $\mathcal{A}$ of L' to the identity. This completes the proof of the assertion and also the proof of Lemma 5.7.

Proof of Theorem 5.6: Let h_t, $0 \leq t \leq 1$, be the isotopy of 5.7. Let $E \subset N$ be an open ball about 0 and let d be the distance from 0 to $R^n - h(E)$. Since $h_t(0) = 0$ and the time interval $0 \leq t \leq 1$ is compact, there exists a small open ball E_1 about 0 with $\bar{E}_1 \subset E$ so that $|h_t(x)| < d$ for all $x \in \bar{E}_1$. Now define

$$\bar{h}_t(x) = \begin{cases} h_t(x) & \text{for } x \in \bar{E}_1 \\ h(x) & \text{for } x \in R^n - E \end{cases}$$

As it stands, this is an isotopy of $h|\bar{E}_1 \cup (R^n - E)$ As an initial step we will extend it to an isotopy of h that satisfies at least the conditions (I) and (II) of Theorem 5.6.

First observe that to <u>any</u> isotopy h_t, $0 \leq t \leq 1$, of h there corresponds a smooth level-preserving imbedding

$$H : [0, 1] \times R^n \longrightarrow [0, 1] \times R^n$$

and conversely. The relation is simply

$$H(t, x) = (t, h_t(x)).$$

The imbedding H determines on its image a vector field

$$\vec{\tau}(t, y) = H(t, x)_* \frac{\partial}{\partial t} = (1, \frac{\partial h_t(x)}{\partial t})$$

where $(t, y) = H(t, x)$, i.e. $y = h_t(x)$. This vector field, together with the imbedding h_o, completely determines h_t and hence H. In fact

$\psi(t, y) = (t, h_t h_o^{-1}(y))$ is the unique family of integral curves with initial values $(0, y) \in 0 \times h_o(R^n)$.

These observations suggest a device due to R. Thom. We will extend the isotopy $\bar{h}_t$ to all of $[0, 1] \times R^n$, by first extending the vector field

$$\vec{\tau}(t, y) = (1, \frac{\partial \bar{h}_t}{\partial t}(h_t^{-1}(y))$$

to a vector field on $[0, 1] \times R^n$ of the form $(1, \vec{\zeta}(t, y))$.

<u>Figure 5.8</u>

<u>The Vector Field</u> $\vec{\tau}(t, y)$

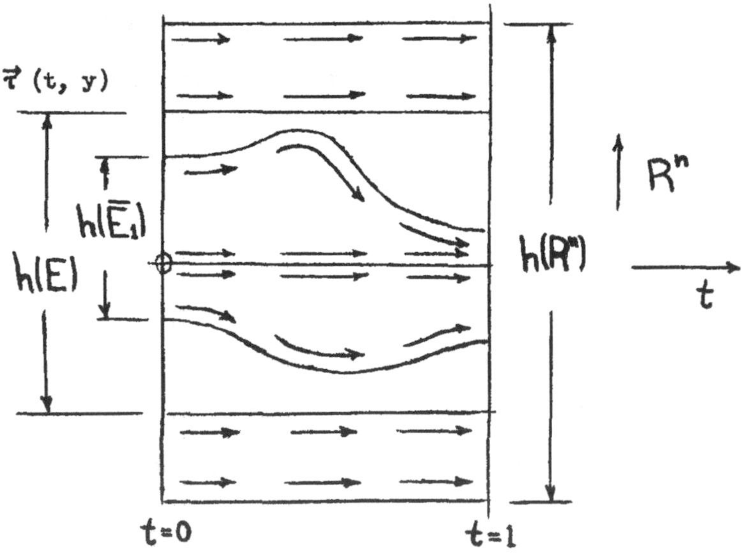

Clearly $\bar{h}_t$ admits an extension to a small open neighborhood of its closed domain $[0, 1] \times \{\bar{E}_1 \cup (R^n - E)\}$. This gives an extension of $\vec{\tau}(t, y)$ to a neighborhood U of its closed domain. Multiplication by a smooth function identically one on the original closed domain and zero outside U produces an extension to $[0, 1] \times R^n$. Finally, setting the first co-ordinate equal to 1 we get a smooth extension

$$\vec{\tau}'(t, y) = (1, \vec{\zeta}(t, y)).$$

Notice that a family of integral curves $\psi(t, y)$ is defined for $y \in R^n$ and for <u>all</u> $t \in [0, 1]$. For $y \in R^n - h(E)$ this is trivial. For $y \in h(E)$ it follows from the fact that the integral curve must remain in the compact set $[0, 1] \times h(\bar{E})$. The family ψ gives a smooth level preserving imbedding

$$\psi : [0, 1] \times R^n \longrightarrow [0, 1] \times R^n$$

Then the equation

$$\psi(t, y) = (t, \bar{h}_t h^{-1}(y))$$

serves to define the required extension of $\bar{h}_t$ to a smooth isotopy of h that satisfies at least conditions (I) and (II) of Theorem 5.6.

Using a similar argument one can prove the following theorem of R. Thom which we will use in Section 8. (For a full proof see Milnor [12, p.5] or Thom [13]).

<u>Theorem 5.8</u> <u>Isotopy Extension Theorem.</u>

Let M be a smooth compact submanifold of the smooth manifold N without boundary. If $h_t, 0 \leq t \leq 1$, is a smooth isotopy of $i : M \subset N$, then h_t is the restriction of a smooth isotopy

$h'_t, 0 \leq t \leq 1$, of the identity map $N \longrightarrow N$ such that h'_t fixes points outside a compact subset of N.

Returning to the proof of Theorem 5.6, let $\bar{h}_t$ denote the extended isotopy. The last condition (III) of Theorem 5.6 will be violated if $\bar{h}_t$ introduces new intersections of the image of R^a with R^b as indicated in Figure 5.9.

Figure 5.9

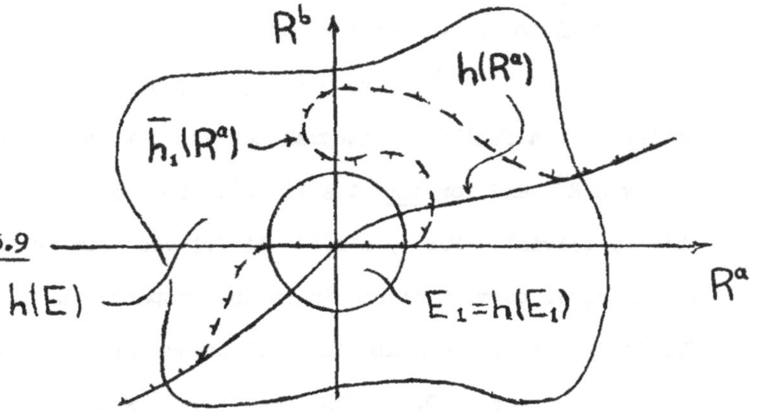

Hence we can use $\bar{h}_t$ only for small values of t, say $t \leq t'$, where no new intersection can occur. We will apply the above process to construct a further deformation of $\bar{h}_{t'}$ which alters $\bar{h}_{t'}$ only at points in E_1, where $\bar{h}_{t'}$ coincides with $h_{t'}$. After a finite number of steps we will obtain the isotopy required. The details follow.

Note that we can write the isotopy h_t of Lemma 5.7 in the form

(*) $$h_t(x) = x_1 h^1(t, x) + \ldots + x_n h^n(t, x)$$

where $h^i(t, x)$ is a smooth function of t and x, $i = 1, \ldots, u$, and (consequently) $h^i(t, 0) = \frac{\partial h_t}{\partial x_i}(0)$. (The proof given in Milnor [4, p.6] is unaffected by the parameter t.)

Lemma 5.9

There exist positive constants K, k such that for all x in a neighborhood of the origin in R^n and all $t \in [0, 1]$

1) $\left|\frac{\partial h_t(x)}{\partial t}\right| < K|x|$

2) $|\pi_a h_t(x)| > k|x|$ for $x \in R^a$, where $\pi_a : R^n \longrightarrow R^a$ is the natural projection.

Proof: The first inequality comes from differentiating (*). The second follows from the fact that $h_t(R^a)$ is transverse to R^b for all t in the compact interval $[0, 1]$.

We now complete the proof of Theorem 5.6 with an inductive step as follows. Suppose we have somehow obtained an imbedding $\tilde{h} : R^n \longrightarrow R^n$ isotopic to h such that

1) For some t_o, $0 \leq t_o \leq 1$, $\tilde{h}(x)$ coincides with $h_{t_o}(x)$ for all x near 0 and with $h(x)$ for all x outside N.

2) $\tilde{h}(R^a) \cap R^b = 0$.

We perform the construction for $\bar{h}_t$ given on pages 61 to 63 taking $\tilde{h}$ in place of h and $[t_o, 1]$ in place of $[0, 1]$ and making the following two special choices (a) and (b).

(a) Choose the ball $E \subset N$ so small that, for all points $x \in E$, $\tilde{h}(x) = h_{t_o}(x)$ and the inequalities of Lemma 5.9 hold.

Note that on the set $[t_o, 1] \times \{\bar{E}_1 \cup (R^n - E)\}$ where $\bar{h}_t$ is initially defined we have

$$(\S) \qquad \left|\frac{\partial h_t(x)}{\partial t}\right| < Kr \text{ , } r = \text{radius of } E.$$

Now $\frac{\partial \bar{h}_t(x)}{\partial t}$ is the R^n-component of $\vec{\tau}(t, y)$. So it is clear from the construction on page (62) that we can

(b) choose the extended R^n-component $\vec{\zeta}(t, y)$ of $\vec{\tau}(t, y)$ to have modulus everywhere less than $k_1 r$.

Then $\bar{h}_t$ will satisfy (§) everywhere in $[t_0, 1] \times R^n$.

We assert that $\bar{h}_t$ will introduce no new intersection of the image of R^a with R^b for $t_0 \leq t \leq t_0 + \frac{k}{K}$. In fact if $x \in R^a \cap (E - E_1)$, the distance of $\bar{h}_{t_0}(x)$ from R^b is

$$|\pi_a \bar{h}_{t_0}(x)| = |\pi_a h_{t_0}(x)| > k\, r$$

Thus (§) shows that for $t_0 \leq t \leq t_0 + \frac{k}{K}$ we have

$$|\pi_a \bar{h}_t(x)| > k\, r - (t - t_0)K\, r \geq 0.$$

Finally, to make possible composition with similar isotopies, we may adjust the parameter t so that the isotopy $\bar{h}_t$, $t_0 \leq t \leq t_0' = \min(1, t_0 + \frac{k}{K})$, satisfies

$$\bar{h}_t(x) = \begin{cases} \tilde{h}(x) & \text{for } t \text{ near } t_o \\ \bar{h}_{t_0'}(x) & \text{for } t \text{ near } t_0' \end{cases}$$

Since the constant k/K depends only on h_t, the required smooth isotopy is a composition of a finite number of isotopies constructed in the above manner. So Theorem 5.6 is complete. This means that Assertion 6) (page 55) is established, and hence the First Cancellation Theorem is proved in general.

§6 A Stronger Cancellation Theorem

Throughout these notes singular homology with integer coefficients will be used unless otherwise specified.

Let M and M' be smooth submanifolds of dimensions r and s in a smooth manifold V of dimension $r + s$ that intersect in points $p_1, \ldots, p_k$, transversely. Suppose that M is oriented and that the normal bundle $\nu(M')$ of M' in V is oriented. At p_i choose a positively oriented r-frame $\xi_1, \ldots, \xi_r$ of linearly independent vectors spanning the tangent space TM_{p_i} of M at p_i. Since the intersection at p_i is transverse, the vectors $\xi_1, \ldots, \xi_r$ represent a basis for the fiber at p_i of the normal bundle $\nu(M')$.

Definition 6.1 The intersection number of M and M' at p_i is defined to be $+1$ or -1 according as the vectors $\xi_1, \ldots, \xi_r$ represent a positively or negatively oriented basis for the fiber at p_i of $\nu(M')$. The intersection number $M' \cdot M$ of M and M' is the sum of the intersection numbers at the points p_i.

Remark 1) In an expression $M' \cdot M$ we agree to write the manifold with oriented normal bundle first.

Remark 2) If V is oriented, any submanifold N is orientable if and only if its normal bundle is orientable. In fact given an orientation for N there is a natural way to give an orientation to $\nu(N)$ and conversely. Namely we require that at any point in N a positively oriented frame tangent to N followed by a frame positively oriented in $\nu(N)$ is a frame positively oriented in V.

Hence if V is oriented there is a natural way to orient $\nu(M)$ and M'. The reader can check that with these orientations.

$$M \cdot M' = (-1)^{rs} M' \cdot M.$$

If the orientation and orientation of normal bundle are not related by the above convention we clearly still have $M \cdot M' = \pm M' \cdot M$ provided V is orientable.

Now assume that M, M' and V are compact connected manifolds without boundary. We prove a lemma which implies that the intersection number $M \cdot M'$ does not change under deformations of M or ambient isotopy of M' and which provides a definition of the intersection number of two closed connected submanifolds of V of complementary dimensions, but not necessarily intersecting transversely. The lemma is based on the following corollary of the Thom Isomorphism Theorem (see the appendix of Milnor [19]) and the Tubular Neighborhood Theorem (see Munkres [5, p.46] and Lang [3, p.73] or Milnor [12, p.19]).

Lemma 6.2 (without proof)

With M' and V as above, there is a natural isomorphism $\psi : H_0(M') \longrightarrow H_r(V, V - M')$.

Let α be the canonical generator of $H_0(M') \cong Z$, and let $[M] \in H_r(M)$ be the orientation generator. The announced lemma is:

<u>Lemma 6.3</u> In the sequence

$$H_r(M) \xrightarrow{g} H_r(V) \xrightarrow{g'} H_r(V, V - M'),$$

where g and g' are induced by inclusion, we have $g' \circ g([M]) = M' \cdot M \, \psi(\alpha)$.

<u>Proof:</u> Choose disjoint open r-cells $U_1, \ldots, U_k$ in M containing $p_1, \ldots, p_k$ respectively. The naturality of the Thom isomorphism implies that the inclusion induced map

$$H_r(U_i, U_i - p_i) \longrightarrow H_r(V, V - M')$$

is an isomorphism given by $\gamma_i \longrightarrow \epsilon_i \, \psi(\alpha)$ where γ_i is the orientation generator of $H_r(U_i, U_i - p_i)$ and ϵ_i is the intersection number of M and M' at p_i. The following commutative diagram, in which the indicated isomorphism comes from excision and the other homomorphisms are induced by inclusion, completes the proof.

$$\begin{array}{ccccc} H_r(M) & \xrightarrow{g} & H_r(V) & \xrightarrow{g'} & H_r(V, V - M') \\ \downarrow & & & \nearrow & \uparrow \\ H_r(M, M - M \cap M') & & \xrightarrow{\cong} & & \sum_{i=1}^{k} H_r(U_i, U_i - p_i) \end{array}$$

We can now reinforce the First Cancellation Theorem 5.4. Let us return to the situation of Theorem 5.4 as set out on page 45. Namely $(W^n; V_0, V_1)$ is a triad with Morse function f having a gradient-like vector field ξ, and p, p' with $f(p) < 1/2 < f(p')$ are the two critical points of f, of index $\lambda, \lambda + 1$ respectively. Suppose that an orientation has been given to the left-hand sphere S_L' in $V = f^{-1}(1/2)$ and also to the normal bundle in V of the right-hand sphere S_R.

<u>Theorem 6.4</u> <u>Second Cancellation Theorem</u>

Suppose W, V_0 and V_1 are simply connected, and $\lambda \geq 2$,

$\lambda + 1 \leq n - 3$. If $S_R \cdot S_L' = \pm 1$, then W^n is diffeomorphic to $V_o \times [0, 1]$. In fact if $S_R \cdot S_L' = \pm 1$, then ξ can be altered near V so that the right- and left-hand spheres in V intersect in a single point, transversely; and the conclusions of Theorem 5.4 then apply.

Remark 1) Observe that $V = f^{-1}(1/2)$ is also simply connected. In fact, applying Van Kampen's theorem (Crowell and Fox [17, p.63]) twice we find $\pi_1(V) \cong \pi_1(D_R^{n-\lambda}(p) \cup V \cup D_L^{\lambda+1}(q))$. (This uses $\lambda \geq 2$, $n - \lambda \geq 3$). But by 3.4 the inclusion $D_R(p) \cup V \cup D_L(q) \subset W$ is a homotopy equivalence. Combining these two statements we see that $\pi_1(V) = 1$.

Remark 2) Notice that conclusion of the theorem is obviously true whenever $\lambda = 0$ or $\lambda = n - 1$. Also the reader can verify with the help of 6.6 below that the theorem holds even with the single dimension restriction $n \geq 6$! (The cases we will not check are $\lambda = 1$ and $\lambda = n - 2$.) The one extension of use to us comes from turning the triad around:

Corollary 6.5 Theorem 6.4 is also valid if the dimension conditions are $\lambda \geq 3$, $(\lambda + 1) \leq n - 2$.

Proof of Corollary: Orient S_R and the normal bundle $\nu S_L'$ of S_L' in V. Now W is simply connected hence orientable. So V is orientable and it follows from Remark 2) page 67 that

$$S_L' \cdot S_R = \pm S_R \cdot S_L' = \pm 1.$$

If we now apply Theorem 6.4 to the triad $(W^n; V_1, V_o)$ with

Morse function $-f$ and gradient-like vector field $-\xi$ we clearly get Corollary 6.5.

The proof of 6.4 will be based on the following delicate theorem which is essentially due to Whitney [7].

Theorem 6.6

Let M and M' be smooth closed, transversely intersecting submanifolds of dimensions r and s in the smooth $(r + s)$-manifold V (without boundary). Suppose that M is oriented and that the normal bundle of M' in V is oriented. Further suppose that $r + s \geq 5$, $s \geq 3$, and, in case $r = 1$ or $r = 2$, suppose that the inclusion induced map $\pi_1(V - M') \longrightarrow \pi_1(V)$ is 1-1 into.

Let $p, q \in M \cap M'$ be points with opposite intersection numbers such that there exists a loop L contractible in V that consists of a smoothly imbedded arc from p to q in M followed by a smoothly imbedded arc from q to p in M' where both arcs miss $M \cap M' - \{p, q\}$.

With these assumptions there exists an isotopy $h_t, 0 \leq t \leq 1$, of the identity $i : V \longrightarrow V$ such that

(i) The isotopy fixes i near $M \cap M' - \{p, q\}$

(ii) $h_1(M) \cap M' = M \cap M' - \{p, q\}$

Figure 6.1

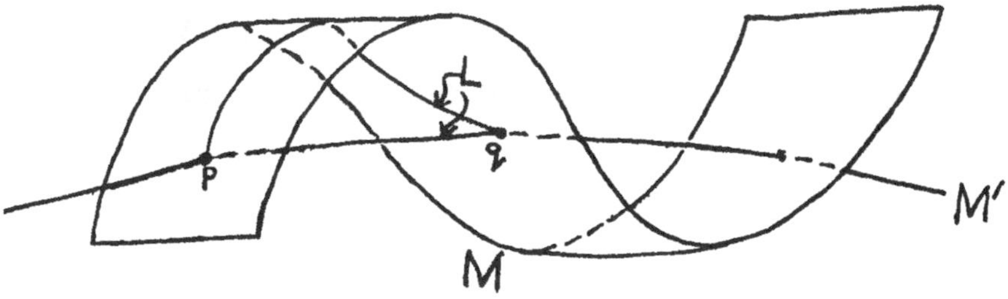

Remark: If M and M' are connected, $r \geq 2$ and V is simply connected no explicit assumption about a loop L need be made. For applying the Hopf-Rinow theorem (see Milnor [4, p.62]), with complete Riemannian metrics on $M - S$ and $M' - S$, where $S = M \cap M' - \{p, q\}$, we can find a smoothly imbedded arc p to q in M and similarly q to p in M' giving a loop L that misses S. The loop L is certainly contractible if V is simply connected.

Proof of Theorem 6.4

According to 5.2 we can make a preliminary adjustment of ξ near V so that S_R and S_L' intersect transversely. If $S_R \cap S_L'$ is not a single point, then $S_R \cdot S_L' = \pm 1$ implies that there exists a pair of points p_1, q_1 in $S_R \cap S_L'$ with opposite intersection numbers. If we can show that Theorem 6.6 applies to this situation, then after we adjust ξ near V, using Lemma 4.7, S_R and S_L' will have two fewer intersection points. Thus if we repeat this process finitely many times S_R and S_L' will intersect transversely in a single point and the proof will be complete.

Since V is simply connected (Remark 1 page 70) it is clear that in case $\lambda \geq 3$ all the conditions of Theorem 6.6 are satisfied. If $\lambda = 2$, it remains to show that $\pi_1(V - S_R) \longrightarrow \pi_1(V) = 1$ is 1-1, i.e. that $\pi_1(V - S_R) = 1$. Now the trajectories of ξ determine a diffeomorphism of $V_0 - S_L$ onto $V - S_R$, where S_L denotes the left-hand 1-sphere of p in V_0. Let N be a product neighborhood of S_L in V_0. Since $n - \lambda - 1 = n - 3 \geq 3$, we have $\pi_1(N - S_L) \cong Z$, and the diagram of fundamental groups corresponding to

$$\begin{array}{ccc} & V_o & \\ \cup & & \cup \\ V_o - S_L & & N \\ \cup & & \cup \\ & (V_o - S_L) \cap N & \\ & = N - S_L & \end{array} \qquad \text{is} \qquad \begin{array}{ccc} & 1 & \\ \nearrow & & \nwarrow \\ \pi_1(V_o - S_L) & & Z \\ \nwarrow & & \nearrow \cong \\ & Z & \end{array}$$

Van Kampen's theorem now implies that $\pi_1(V_o - S_L) = 1$. This completes the proof of Theorem 6.4 modulo proving Theorem 6.6.

Proof of 6.6

Suppose that the intersection numbers at p and q are $+1$ and -1 respectively. Let C and C' be the smoothly imbedded arcs in M and M' from p to q extended a little way at both ends. Let C_o and C_o' be open arcs in the plane intersecting transversely in points a and b, and enclosing a disk D (with two corners) as in Figure 6.2 below. Choose an embedding $\varphi_1 : C_o \cup C_o' \longrightarrow M \cup M'$ so that $\varphi_1(C_o)$ and $\varphi_1(C_o')$ are the arcs C and C', with a and b corresponding to p and q. The theorem will follow quickly from the next lemma, which embeds a standard model.

Lemma 6.7 For some neighborhood U of D we can extend $\varphi_1 | U \cap (C_o \cup C_o')$ to an embedding $\varphi : U \times R^{r-1} \times R^{s-1} \longrightarrow V$ such that $\varphi^{-1}(M) = (U \cap C_o) \times R^{r-1} \times 0$ and $\varphi^{-1}(M') = (U \cap C_o') \times 0 \times R^{s-1}$.

Figure 6.2 The Standard Model

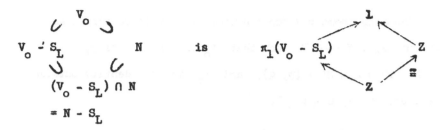

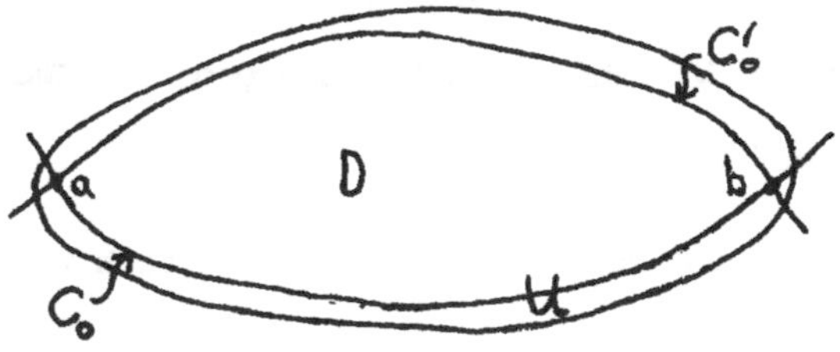

Assuming Lemma 6.7 for the moment, we will construct an isotopy $F_t : V \longrightarrow V$ such that F_0 is the identity, $F_1(M) \cap M' = M \cap M' - \{p, q\}$, and F_t is the identity outside the image of φ, $0 \leq t \leq 1$.

Let W denote $\varphi(U \times R^{r-1} \times R^{s-1})$ and define F_t to be the identity on $V - W$. Define F_t on W as follows.

Choose an isotopy $G_t : U \longrightarrow U$ of our plane model such that

1.) G_0 is the identity map,

2.) G_t is the identity in a neighborhood of the boundary $\bar{U} - U$ of U, $0 \leq t \leq 1$, and

3.) $G_1(U \cap C_0) \cap C_0' = \emptyset$. (See Figure 6.3.)

Figure 6.3

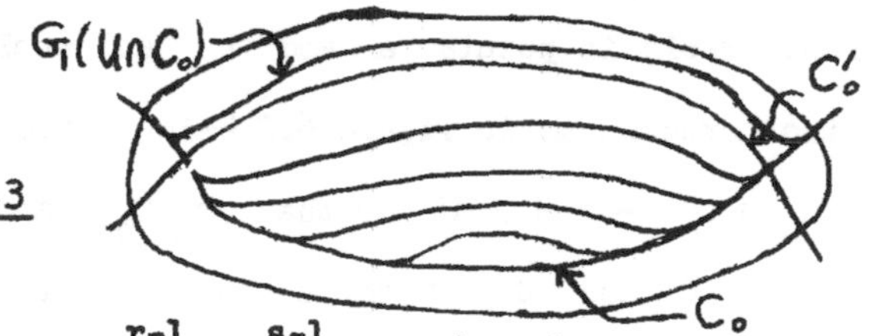

Let $\rho : R^{r-1} \times R^{s-1} \longrightarrow [0, 1]$ be a smooth function such that with $x \in R^{r-1}$, $y \in R^{s-1}$

$$\rho(x, y) = \begin{cases} 1 & \text{for } |x|^2 + |y|^2 \leq 1 \\ 0 & \text{for } |x|^2 + |y|^2 \geq 2. \end{cases}$$

Define an isotopy $H_t : U \times R^{r-1} \times R^{s-1} \longrightarrow U \times R^{r-1} \times R^{s-1}$ by

$$H_t(u, x, y) = (G_{t\rho(x, y)}(u), x, y) , \ u \in U.$$

It is easy to see that $F_t(w) = \varphi \circ H_t \circ \varphi^{-1}(w)$, $w \in W$, defines the required isotopy on W. This finishes the proof of Theorem 6.6, modulo proving 6.7.

Lemma 6.8 There exists a Riemannian metric on V such that

1.) in the associated connection (see Milnor [4, p.44]) M and M' are totally geodesic submanifolds of V (i.e. if a geodesic in V is tangent to M or to M' at any point then it lies entirely in M or M', respectively.)

2.) there exist coordinate neighborhoods N_p and N_q about p and q in which the metric is the euclidean metric and so that $N_p \cap C$, $N_p \cap C'$, $N_q \cap C$, and $N_q \cap C'$ are straight line segments.

<u>Proof</u> (due to E. Feldman): We know that M intersects M' transversely in points $p_1, \ldots, p_k$ with $p = p_1$ and $q = p_2$. Cover $M \cup M'$ by coordinate neighborhoods $W_1, \ldots, W_m$ in V with coordinate diffeomorphisms $h_i : W_i \longrightarrow R^{r+s}$, $i = 1, \ldots, m$, such that

a.) there are disjoint coordinate neighborhoods $N_1, \ldots, N_k$ with $p_i \in N_i \subset \bar{N}_i \subset W_i$ and $N_i \cap W_j = \emptyset$ for $i = 1, \ldots, k$ and $j = k + 1, \ldots, m$.

b.) $h_i\,(W_i \cap M) \subset R^r \times 0$
$h_i\,(W_i \cap M') \subset 0 \times R^s$ $\qquad i = 1, \ldots, k.$

c.) $h_i\,(W_i \cap C)$ and $h_i\,(W_i \cap C')$ are straight line segments in R^{r+s}, $i = 1,2$.

Construct a Riemann metric $< \vec{v}, \vec{w} >$ on the open set $W_0 = W_1 \cup \ldots \cup W_m$ by piecing together the metrics on the W_i induced by the h_i, $i = 1, \ldots, m$, using a partition of unity. Note that because of a.) this metric is euclidean in the N_i, $i = 1, \ldots, k$.

With this metric construct open tubular neighborhoods T and T' of M and M' in W_o using the exponential map (see Lang [3, p. 73]). By choosing them thin enough we may assume that $T \cap T' \subset N_1 \cup \ldots \cup N_k$ and that

$$h_i(T \cap T' \cap N_i) = OD^r_\epsilon \times OD^s_{\epsilon'} \subset R^r \times R^s = R^{r+s} ,$$

$i = 1, \ldots, k$, for some ϵ, $\epsilon' > 0$ depending on i. The situation is represented schematically in Figure 6.4.

Figure 6.4

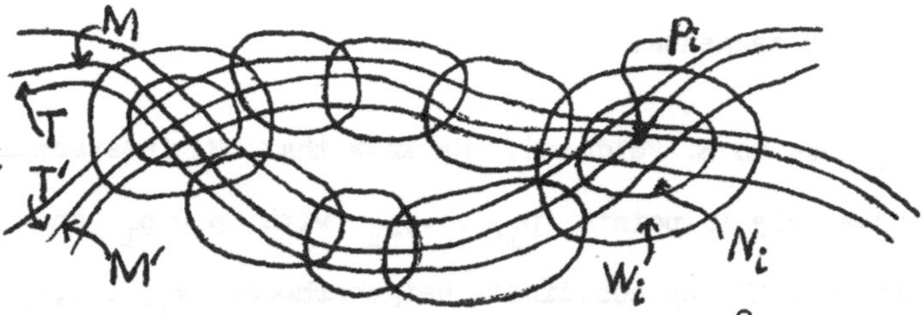

Let $A : T \longrightarrow T$ be the smooth involution ($A^2 = A \circ A =$ identity) which is the antipodal map on each fiber of T. Define a new Riemann metric $< \vec{v}, \vec{w}>_A$ on T by $< \vec{v}, \vec{w}>_A = \frac{1}{2} (< \vec{v}, \vec{w}> + < A_*\vec{v}, A_*\vec{w}>)$.

Assertion: With respect to this new metric, M is a totally geodesic submanifold of T. To see this, let ω be a geodesic in T tangent to M at some point $z \in M$. It is easy to see that A is an isometry of T in the new metric and hence sends geodesics to geodesics. Since M is the fixed point set of A, it follows that $A(\omega)$ and ω are geodesics with the same tangent vector at $A(Z) = Z$. By uniqueness of geodesics, A is the identity on ω. Therefore $\omega \subset M$, which proves the assertion.

Similarly define a new metric $< \vec{v}, \vec{w}>_{A'}$ on T'. It follows from property b.) and the form of $T \cap T'$ that these two new metrics agree with the old metric on $T \cap T'$ and hence together define a metric on $T \cup T'$. Extending to all of V the restriction of this metric to an open set O, with $M \cup M' \subset O \subset \bar{O} \subset T \cup T'$, completes the construction of a metric on V satisfying conditions 1.) and 2.).

Proof of Lemma 6.7 (The proof occupies the rest of Section 6)

Choose a Riemannian metric on V provided by Lemma 6.8. Let $\tau(p)$, $\tau(q)$, $\tau'(p)$, $\tau'(q)$ be the unit vectors tangent to the arcs C and C' (oriented from p to q) at p and q. Since C is a contractible space, the bundle over it of vectors orthogonal to M is trivial. Using this fact construct a field of unit vectors along C orthogonal to M and equal to the parallel translates of $\tau'(p)$ and of $-\tau'(q)$ along $N_p \cap C$ and $N_q \cap C$ respectively.

Construct some corresponding vector field in the model. (See Figure 6.5)

Figure 6.5

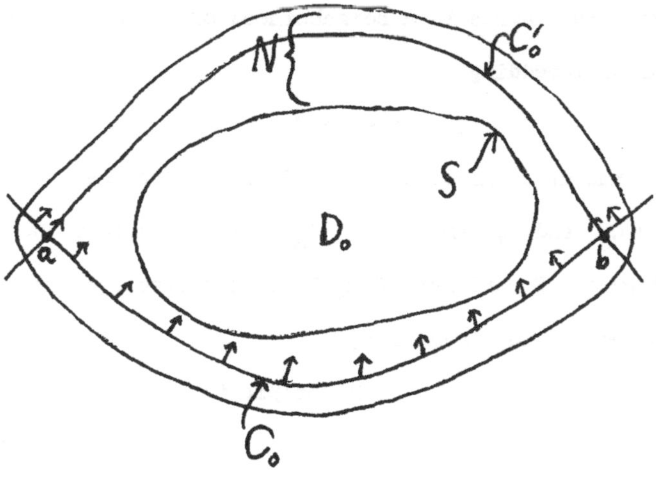

Using the exponential map, we see that there exists a neighborhood of C_o in the plane and an extension of $\varphi_1|C_o$ to an imbedding of this neighborhood into V. Actually, the expnnential map gives an imbedding locally, and then one uses the following lemma, whose elementary proof may be found in Munkres [5, p.49 Lemma 5.7 (which is incorrectly stated)]

<u>Lemma 6.9</u> Let A_o be a closed subset of a compact metric space A. Let $f : A \longrightarrow B$ be a local homeomorphism such that $f|A_o$ is 1-1. Then there is a neighborhood of W of A_o such that $f|W$ is 1-1.

Similarly extend $\varphi_1|C_o'$ to an imbedding of a neighborhood of C_o' using a field of unit vectors along C' orthogonal to M' which along $N_p \cap C'$ and $N_q \cap C'$ consists of the parallel translates of $\tau(p)$ and $-\tau(q)$ respectively. When $r = 1$ this is possible only because the intersection numbers at p and q are opposite.

Using property 2) of the metric on V (see 6.8) we see that the imbeddings agree in a neighborhood of $C_o \cup C_o'$ and hence define an imbedding

$$\varphi_2 : N \longrightarrow V$$

of a closed annular neighborhood N of BdD such that $\varphi_2^{-1}(M) = N \cap C_o$ and $\varphi_2^{-1}(M') = N \cap C_o'$. Let S denote the inner boundary of N and let $D_o \subset D$ be the disc bounded by S in the plane. (See Figure 6.5)

Since the given loop L is homotopic to the loop $\varphi_2(S)$, the latter is contractible in V. Actually $\varphi_2(S)$ is contractible in $V - (M \cup M')$ as the following lemma will show.

Lemma 6.10

If V_1^n, $n \geq 5$, is a smooth manifold, M_1 a smooth submanifold of codimension at least 3, then a loop in $V_1 - M_1$ that is contractible in V_1 is also contractible in $V_1 - M_1$.

Before proving 6.10 we recall two theorems of Whitney.

Lemma 6.11 (See Milnor [15, p.62 and p.63]) Let $f : M_1 \longrightarrow M_2$ be a continuous map of smooth manifolds which is smooth on a closed subset A of M_1. Then there exists a smooth map $g : M_1 \longrightarrow M_2$ such that $g \simeq f$ (g is homotopic to f) and $g|A = f|A$.

Lemma 6.12 (See Whitney [16] and Milnor [15, p.63])

Let $f : M_1 \longrightarrow M_2$ be a smooth map of smooth manifolds which is an imbedding on the closed subset A of M_1. Assume that $\dim M_2 \geq 2 \dim M_1 + 1$. Then there exists an imbedding $g : M_1 \longrightarrow M_2$ approximating f such that $g \simeq f$ and $g|A = f|A$.

Proof of 6.10:

Let $g : (D^2, S^1) \longrightarrow (V_1, V_1 - M_1)$ give a contraction in V_1 of a loop in $V_1 - M_1$. Because $\dim (V_1 - M_1) \geq 5$ the above lemmas give a smooth imbedding

$$h : (D^2, S^1) \longrightarrow (V_1, V_1 - M_1)$$

such that $g_{|S^1}$ is homotopic to $h_{|S^1}$ in $V_1 - M_1$.

The normal bundle of $h(D^2)$ is trivial since $h(D^2)$ is contractible. Hence there exists an imbedding H of $D^2 \times R^{n-2}$

into V_1 such that $H(u, o) = h(u)$ for $u \in D^2$. Take $\varepsilon > 0$ so small that $|\vec{x}| < \epsilon$, $\vec{x} \in R^{n-2}$, implies $H(S^1 \times \vec{x}) \subset V_1 - M_1$. Since codimension $M_1 \geq 3$, there exists (cf. 4.6) $x_o \in R^{n-2}$, $|\vec{x}_o| < \epsilon$, such that $H(D^2 \times \vec{x}_o) \cap M_1 = \emptyset$. Now in $V_1 - M_1$ we have $g_{|S^1} \simeq h_{|S^1} = H_{|S^1 \times 0} \simeq H_{|S^1 \times \vec{x}_o} \simeq$ constant. This completes the proof of Lemma 6.10.

Now we can show that $\varphi(S)$ is contractible in $V - M \cup M'$. For it is contractible in $V - M'$ by 6.10 if $r \geq 3$, and if $r = 2$ by the hypothesis that $\pi_1(V - M') \longrightarrow \pi_1(V)$ is 1-1. Then, since $s \geq 3$, $\varphi(S)$ is also contractible in $(V - M') - M = V - (M \cup M')$ by Lemma 6.10.

We now choose a continuous extension of φ_2 to $U = N \cup D_o$

$$\varphi_2' : U \longrightarrow V$$

that maps Int D into $V - (M \cup M')$. Applying Lemmas 6.11 and 6.12 to $\varphi_2'|\text{Int } D$ we can obtain a smooth imbedding $\varphi_3 : U \longrightarrow V$ coinciding with φ_2 on a neighborhood of $U - \text{Int } D$, and such that $\varphi_3(u) \notin M \cup M'$ for $u \notin C_o \cup C_o'$.

It remains now to extend φ_3 to $U \times R^{r-1} \times R^{s-1}$ as desired.

We let U' denote $\varphi_3(U)$, and for convenience in notation we shall write C, C', C_o, and C_o' in place of $U' \cap C$, $U' \cap C'$, $U \cap C_o$, and $U \cap C_o'$, respectively.

<u>Lemma 6.13</u> There exist smooth vector fields $\xi_1, \ldots, \xi_{r-1}$, $\eta_1, \ldots, \eta_{s-1}$ along U' which satisfy condition 1.) below and such that $\xi_1, \ldots, \xi_{r-1}$ satisfy 2.) and $\eta_1, \ldots, \eta_{s-1}$ satisfy 3.)

1.) <u>are orthonormal and are orthogonal to</u> U'

2.) <u>along</u> C <u>are tangent</u> M

3.) <u>along</u> C' <u>are tangent to</u> M'.

<u>Proof</u>: The idea is to construct $\xi_1, \ldots, \xi_{r-1}$ in steps, first along C by parallel translation, then extending to $C \cup C'$ by a bundle argument, and then to U' by another bundle argument. The details follow.

Let τ and τ' be the normalized velocity vectors along C and C', and let ν' be the field of unit vectors along C' which are tangent to U' and are inward orthogonal to C'. Then $\nu'(p) = \tau(p)$ and $\nu'(q) = -\tau(q)$ (see Figure 6.6)

<u>Figure 6.6</u>

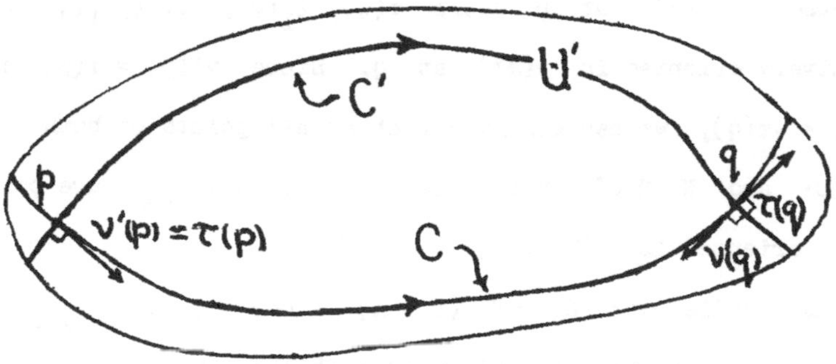

Choose $r - 1$ vectors $\xi_1(p), \ldots, \xi_{r-1}(p)$ which are tangent to M at p, are othogonal to U', and are such that the r-frame $\tau(p), \xi_1(p), \ldots, \xi_{r-1}(p)$ is positively oriented in TM_p. Parallel translating these $r - 1$ vectors along C gives $r - 1$ smooth vector fields $\xi_1, \ldots, \xi_{r-1}$ along C. These vectors fields satisfy 1.) because parallel translation preserves inner products (see Milnor [4, p.48]). They satisfy 2.) because parallel translation along a curve in a totally geodesic submanifold M sends tangent vectors to M into tangent vectors to M (see Helgason, Differential Geometry and Symmetric Spaces, p. 80). Actually, given the construction of the Riemannian metric in 6.8, condition 2.) easily follows from the existence of the "antipodal isometry" A on a tubular neighborhood of M (compare the argument on p.76). Finally, by continuity the r-frame $\tau, \xi_1, \ldots, \xi_{r-1}$ is positively oriented in TM (= tangent bundle of M) at every point of C.

Now parallel translate $\xi_1(p), \ldots, \xi_{r-1}(p)$ along $N_p \cap C'$ and $\xi_1(q), \ldots, \xi_{r-1}(q)$ along $N_q \cap C'$. By hypothesis the intersection numbers of M and M' at p and q are $+1$ and -1. This means that $\tau(p), \xi_1(p), \ldots, \xi_{r-1}(p)$ is positively oriented in $\nu(M')$ at p while $\tau(q), \xi_1(q), \ldots, \xi_{r-1}(q)$ is negatively oriented in $\nu(M')$ at q. Since $\nu'(p) = \tau(p)$ and $\nu'(q) = -\tau(q)$, we can conclude that at all points of both $N_p \cap C'$ and $N_q \cap C'$, the frames $\nu', \xi_1, \ldots, \xi_{r-1}$ are positively oriented in $\nu(M')$.

The bundle over C' of $(r-1)$-frames $\zeta_1, \ldots, \zeta_{r-1}$ orthogonal to M' and to U', and such that $\nu', \zeta_1, \ldots, \zeta_{r-1}$ is positively oriented in $\dot{\nu}(M')$ is trivial with fiber $SO(r-1)$, which is connected. Hence we may extend $\xi_1, \ldots, \xi_{r-1}$ to a smooth field of $(r-1)$-frames on $C \cup C'$ that satisfy conditions 1) and 2).

The bundle over U' of orthonormal $(r-1)$-frames orthogonal to U' is a trivial bundle with fiber $O(r+s-2)/O(s-1) = V_{r-1}(R^{r+s-2})$, the Stiefel manifold of orthonormal $(r-1)$-frames in R^{r+s-2}. So far we have constructed a smooth cross-section $\xi_1, \ldots, \xi_{r-1}$ of this bundle over $C \cup C'$. Composing $\xi_1, \ldots, \xi_{r-1}$ with the projection into the fiber, we get a smooth map of $C \cup C'$ into $O(r + s - 2)/O(s - 1)$ which is simply connected since $s \geq 3$. (see Steenrod [18, p.103]). Hence there is a continuous extension to U' and by Lemma 6.11 there exists a smooth extension. Thus we can define $\xi_1, \ldots, \xi_{r-1}$ over all of U' to satisfy 1.) and 2.).

To define the remaing desired vector fields, observe that the bundle over U' of orthonormal frames $\eta_1, \ldots, \eta_{s-1}$ in TV such that each η_i is orthogonal to U' and to $\xi_1, \ldots, \xi_{r-1}$ is a trivial bundle because U' is contractible. Let the desired field of frames $\eta_1, \ldots, \eta_{s-1}$ on U' be a smooth cross-section of this bundle. Then $\xi_1, \ldots, \xi_{r-1}, \eta_1, \ldots, \eta_{s-1}$ satisfy 1.). Furthermore, since $\xi_1, \ldots, \xi_{r-1}$ are orthogonal to M' along C', it follows that $\eta_1, \ldots, \eta_{s-1}$ satisfy 3.). This finishes the proof of Lemma 6.13.

Completion of Proof of Lemma 6.7

Define a map $U \times R^{r-1} \times R^{s-1} \longrightarrow V$ by

$$(u, x_1, \ldots, x_{r-1}, y_1, \ldots, y_{s-1}) \longrightarrow \exp\Big[\sum_{i=1}^{r-1} x_i \xi_i(\varphi_3(u)) + \sum_{j=1}^{s-1} y_j \eta_j(\varphi_3(u))\Big].$$

It follows from Lemma 6.9 and the fact that this map is a local diffeomorphism that there exists an open ϵ-neighborhood N_ϵ about the origin in $R^{r+s-2} = R^{r-1} \times R^{s-1}$ such that if $\varphi_4 : U \times N_\epsilon \longrightarrow V$ denotes this map restricted to $U \times N_\epsilon$ then

φ_4 is an embedding. (U may have to be replaced by a slightly smaller neighborhood, which we still denote by U.)

Define an embedding $\varphi : U \times R^{r-1} \times R^{s-1} \longrightarrow V$ by $\varphi(u, z) = \varphi_4(u, \frac{\epsilon z}{\sqrt{1+|z|^2}})$. Then $\varphi(C_o \times R^{r-1} \times 0) \subset M$ and $\varphi(C_o' \times 0 \times R^{s-1}) \subset M'$ because M and M' are totally geodesic submanifolds of V. Moreover, since $\varphi(U \times 0) = U'$ intersects M and M' precisely in C and C', transversely, it follows that, for $\epsilon > 0$ sufficiently small, Image(φ) intersects M and M' precisely in the above product neighborhoods of C and C'. This means $\varphi^{-1}(M) = C_o \times R^{r-1} \times 0$ and $\varphi^{-1}(M') = C_o' \times 0 \times R^{s-1}$. Thus φ is the required embedding. This ends the proof of Lemma 6.7.

§7 Cancellation of Critical Points in the Middle Dimensions

Definition 7.1 Suppose W is a compact oriented smooth n-dimensional manifold, and set $X = \mathrm{Bd}W$. It is easy to check that X is given a well-defined orientation, called the induced orientation, by saying that an $(n - 1)$-frame $\tau_1, \ldots, \tau_{n-1}$ of vectors tangent to X at some point $x \in X$ is positively oriented if the n-frame $v, \tau_1, \ldots, \tau_{n-1}$ is positively oriented in TW_x, where v is any vector at x tangent to W but not to X and pointing out of W (i.e. v is outward normal to X).

Alternatively, one specifies $[X] \in H_{n-1}(X)$ as the induced orientation generator for X, where $[X]$ is the image of the orientation generator $[W] \in H_n(W, X)$ for W under the boundary homomorphism $H_n(W, X) \longrightarrow H_{n-1}(X)$ of the exact sequence for the pair (W, X).

Remark: The reader can easily give a natural correspondence between an orientation of a compact manifold M^n specified by an orientation of the tangent bundle (in terms of ordered frames) and an orientation M specified by a generator $[M]$ of $H_n(M; Z)$ (cf. Milnor [19, p.21]). It is not difficult to see that the two ways given above to orient BdW are equivalent under this natural correspondence. Since we will always use the second way to orient BdW, the proof is omitted.

Suppose now that we are given n-dimensional triads $(W; V, V')$, $(W'; V', V'')$, and $(W \cup W'; V, V'')$. Suppose also that f is a Morse function on $W \cup W'$ with critical points $q_1, \ldots, q_\ell \in W$ and $q'_1, \ldots, q'_m \in W'$ such that $q_1, \ldots, q_\ell$ are all on one level and are of index λ, while $q'_1, \ldots, q'_m$ are on another level and are of index $\lambda + 1$, and V' is a non-critical level between them. Choose a gradient-like vector field for f and orient the left-hand disks $D_L(q_1), \ldots, D_L(q_\ell)$ in W and $D'_L(q'_1), \ldots, D'_L(q'_m)$ in W'.

Orientation for the normal bundle $\nu D_R(q_i)$ of a right-hand disk in W is then determined by the condition that $D_L(q_i)$ have intersection number +1 with $D_R(q_i)$ at the point q_i. The normal bundle $\nu S_R(q_i)$ of $S_R(q_i)$ in V' is naturally isomorphic to the restriction of $\nu D_R(q_i)$ to $S_R(q_i)$. Hence the orientation of $\nu D_R(q_i)$ determines an orientation for $\nu S_R(q_i)$.

Combining Definition 7.1 and the above paragraph we conclude that once orientations have been chosen for the left-hand disks in W and W', there is a natural way to orient the left-hand spheres in V' and the normal bundles of the right-hand spheres in V'. Consequently the intersection number $S_R(q_i) \cdot S'_L(q'_j)$ of left-hand spheres with right-hand spheres in V' are well defined.

From Section 3 we know that $H_\lambda(W, V)$ and $H_{\lambda+1}(W \cup W', W) \cong H_{\lambda+1}(W', V')$ are free abelian with generators $[D_L(q_1)], \ldots, [D_L(q_\ell)]$ and $[D'_L(q'_1)], \ldots, [D'_L(q'_m)]$, respectively, represented by the oriented left-hand disks.

<u>Lemma 7.2</u> Let M be an oriented closed smooth manifold of dimension λ embedded in V' with $[M] \in H_\lambda(M)$ the orientation

generator, and let $h : H_\lambda(M) \longrightarrow H_\lambda(W, V)$ be the map induced by inclusion. Then $h([M]) = S_R(q_1)\cdot M\ [D_L(q_1)] + \ldots + S_R(q_\ell)\cdot M\ [D_L(q_\ell)]$ where $S_R(q_i)\cdot M$ denotes the intersection number of $S_R(q_i)$ and M in V'.

<u>Corollary 7.3</u> With respect to the bases represented by the oriented left-hand disks, the boundary map $\partial : H_{\lambda+1}(W \cup W', W) \longrightarrow H_\lambda(W, V)$ for the triple $W \cup W' \supset W \supset V$ is given by the matrix (a_{ij}) of intersection numbers $a_{ij} = S_R(q_i) \cdot S'_L(q'_j)$ in V', naturally determined by the orientations assigned to the left-hand disks.

<u>Proof of Corollary</u>: Consider one of the basis elements $[D'_L(q'_j)] \in H_{\lambda+1}(W \cup W', W)$. We can factor the map ∂ into the composition

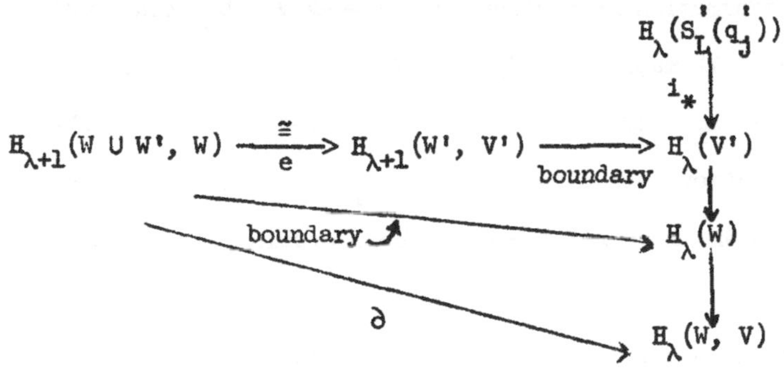

Here e is the inverse of the excision isomorphism, and i_* is induced by inclusion.

By definition of the orientation for $S'_L(q'_j)$, we have

$$\text{boundary} \circ e([D'_L(q'_j)]) = i_*([S'_L(q'_j)]).$$

The result follows by setting $M = S'_L(q'_j)$ in Lemma 7.2.

Proof of 7.2: We assume $\ell = 1$, the proof in the general case being similar. Set $q = q_1$, $D_L = D_L(q_1)$, $D_R = D_R(q_1)$, and $S_R = S_R(q_1)$. We must show that $h([M]) = S_R \cdot M \cdot [D_L]$.

Consider the following diagram:

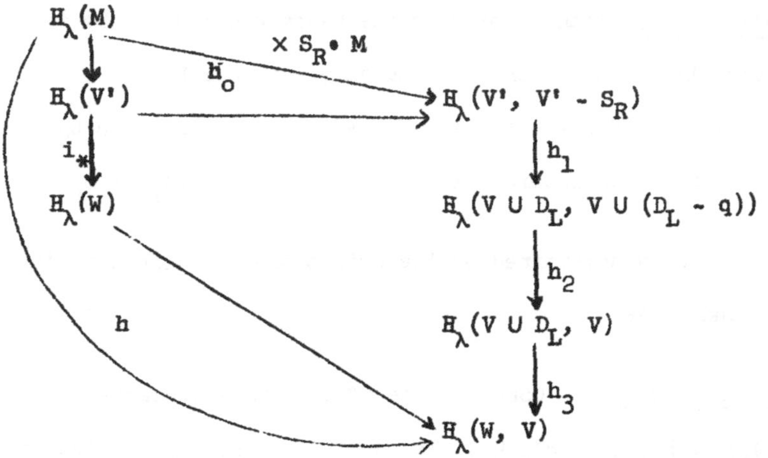

The deformation retraction $r : W \longrightarrow V \cup D_L$ constructed in Theorem 3.14 maps $V' - S_R$ to $V \cup (D_L - q)$, so the homomorphism h_1 induced by $r|V'$ is well-defined. The obvious deformation retraction of $V \cup (D_L - q)$ to V induces the isomorphism h_2. All the other homomorphisms are induced by inclusion.

The diagram commutes because $i_* = (r|V')_*$ (since the maps $i, r|V' : V' \longrightarrow W$ are homotopic) and the corresponding diagram of topological spaces and continuous maps commutes pointwise, with $r|V'$ in place of i.

From Lemma 6.3 we know that $h_0([M]) = S_R \cdot M \;\; \psi(\alpha)$ where $\alpha \in H_0(S_R)$ is the canonical generator and $\psi : H_0(S_R) \longrightarrow H_\lambda(V', V' - S_R)$ is the Thom isomorphism. Hence in order to prove that $h([M]) = S_R \cdot M\ [D_L]$, by commutativity of the diagram it suffices to show that

(*) $$h_3 \circ h_2 \circ h_1(\psi(\alpha)) = [D_L].$$

The class $\psi(\alpha)$ is represented by any oriented disk D^λ which intersects S_R in one point x, transversely, with intersection number $S_R \cdot D^\lambda = +1$. Referring to the standard form for an elementary cobordism given in Theorem 3.13, and the conventions by which $D(S_R)$ is oriented, one can see that the image $r(D^\lambda)$ of D^λ under the retraction r represents

$$D_R \cdot D^\lambda = S_R \cdot D^\lambda = +1$$

times the orientation generator $h_2^{-1}h_3^{-1}([D_L])$ for $H_\lambda(V \cup D_L, V \cup (D_L - q))$. It follows that

$$h_1\psi(\alpha) = h_2^{-1}h_3^{-1}([D_L])$$

or

$$h_3h_2h_1\ \psi(\alpha) = [D_L]$$

as required. Thus the proof of 7.2 is complete.

Given any cobordism c represented by the triad $(W; V, V')$, according to 4.8 we can factor $c = c_0 c_1 \dots c_n$ so that c_λ admits a Morse function all of whose critical points are on the same level and have index λ. Let $c_0 c_1 \dots c_\lambda$ be represented by the manifold $W_\lambda \subset W$, $\lambda = 0, 1, \dots, n$, and set $W_{-1} = V$, so that

$$V = W_{-1} \subset W_0 \subset W_1 \subset \dots \subset W_n = W$$

Define $C_\lambda = H_\lambda(W_\lambda, W_{\lambda-1}) \cong H_*(W_\lambda, W_{\lambda-1})$ and let $\partial : C_\lambda \longrightarrow C_{\lambda-1}$ be the boundary homomorphism for the exact sequence of the triple $W_{\lambda-2} \subset W_{\lambda-1} \subset W_\lambda$.

<u>Theorem 7.4</u> $C_* = \{C_\lambda, \partial\}$ is a chain complex (i.e. $\partial^2 = 0$) and $H_\lambda(C_*) \cong H_\lambda(W, V)$ for all λ.

<u>Proof</u>: (Note that we do not use the fact that C_λ is free abelian, but only that $H_*(W_\lambda, W_{\lambda-1})$ is concentrated in degree λ.)

That $\partial^2 = 0$ is clear from the definition. To prove the isomorphism, consider the following diagram.

$$\begin{array}{ccccccc}
 & & 0 & & & & \\
 & & \downarrow & & & & \\
H_{\lambda+1}(W_{\lambda+1}, W_\lambda) & \longrightarrow & H_\lambda(W_\lambda, W_{\lambda-2}) & \longrightarrow & H_\lambda(W_{\lambda+1}, W_{\lambda-2}) & \longrightarrow & 0 \\
\| & \overset{\partial}{\searrow} & \downarrow & \nearrow & & & \\
C_{\lambda+1} & & H_\lambda(W_\lambda, W_{\lambda-1}) = C_\lambda & & & & \\
 & & \partial \downarrow & & & & \\
 & & H_{\lambda-1}(W_{\lambda-1}, W_{\lambda-2}) = C_{\lambda-1} & & & &
\end{array}$$

The horizontal is the exact sequence of the triple $(W_{\lambda+1}, W_\lambda, W_{\lambda-2})$ and the vertical is the exact sequence of $(W_\lambda, W_{\lambda-1}, W_{\lambda-2})$. One checks easily that the diagram commutes. Then clearly $H_\lambda(C_*) \cong H_\lambda\ (W_{\lambda+1}, W_{\lambda-2})$. But $H_\lambda(W_{\lambda+1}, W_{\lambda-2}) \cong H_\lambda(W, V)$. Leaving the reader to verify this last statement (see Milnor [19, p.9]), we have the desired isomorphism $H_\lambda(C_*) \cong H_\lambda(W, V)$.

<u>Theorem 7.5</u> (<u>Poincare Duality</u>.)

If $(W; V, V')$ is a smooth manifold triad of dimension n and W is oriented, then $H_\lambda(W, V)$ is isomorphic to $H^{n-\lambda}(W, V')$ for all λ.

<u>Proof</u>: Let $c = c_0 c_1 \ldots c_n$ and $C_* = \{C_\lambda, \partial\}$ be defined with respect to a Morse function f as above, and fix a gradient-like

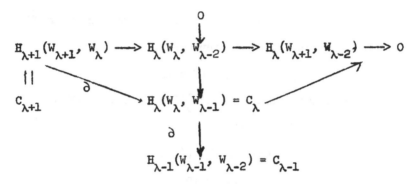

vector field ξ for f. Given fixed orientations, the left-hand disks of c_λ form a basis for $C_\lambda = H_\lambda(W_\lambda, W_{\lambda-1})$. From 7.3 we know that with respect to this basis the boundary map $\partial : C_\lambda \longrightarrow C_{\lambda-1}$ is given by the matrix of intersection numbers of oriented left-hand spheres of c_λ with right-hand spheres of $c_{\lambda-1}$ having oriented normal bundles.

Similarly let $W'_\mu \subset W$ represent $c_{n-\mu}\ c_{n-\mu+1} \cdots c_n$ for $\mu = 0, 1, \ldots, n$ and set $W'_{-1} = V'$. Define $C'_\mu = H_\mu(W'_\mu, W'_{\mu-1})$ and $\partial' : C'_\mu \longrightarrow C'_{\mu-1}$ as before. For any right-hand disk D_R, the given orientation of $\nu(D_R)$ (from the oriented left-hand disk) together with the orientation of W give a naturally defined orientation for D_R. Then $\partial : C'_\mu \longrightarrow C'_{\mu-1}$ is given by a matrix of intersection number of oriented right-hand spheres with left-hand spheres having oriented normal bundles.

Let $C'^* = \{C'^\mu, \delta'\}$ be the cochain complex dual to the chain complex $C'_* = \{C'_\mu, \partial'\}$ (Thus $C'^\mu = \mathrm{Hom}(C'_\mu, Z)$). Choose as basis for C'^μ the basis dual to the basis of C'_μ which is determined by the oriented right-hand disks of $c_{n-\mu}$.

An isomorphism $C_\lambda \longrightarrow C'^{n-\lambda}$ is induced by assigning to each oriented left-hand disk, the dual of the oriented right-hand disk of the same critical point. Now, as we have stated, $\partial : C_\lambda \longrightarrow C_{\lambda-1}$ is given by a matrix $(a_{ij}) = (S_R(p_i) \cdot S'_L(p'_j))$.

It is easy to see that $\delta' : C'^{n-\lambda} \longrightarrow C'^{n-\lambda+1}$ is given by the matrix $(b_{ij}) = (S'_L(p'_j) \cdot S_R(p_i))$. But since W is oriented,

$b_{ij} = \pm a_{ij}$, the sign depending only on λ. (cf. 6.1 Remark 2. The sign turns out to be $(-1)^{\lambda-1}$.) Thus ∂ corresponds to $\pm \delta'$, and it follows that the isomorphism of chain groups induces an isomorphism $H_\lambda(C_*) \cong H^{n-\lambda}(C'^*)$.

Now 7.1 implies $H_\lambda(C_*) \cong H_\lambda(W, V)$ and $H_\mu(C'_*) \cong H_\mu(W, V')$ for each λ and μ. Moreover, the latter isomorphism implies that $H^\mu(C'^*) \cong H^\mu(W, V')$ for each μ. For if two chain complexes have isomorphic homology then the dual cochain complexes have isomorphic cohomology. This follows from the Universal Coefficient Theorem.

Combining the last two paragraphs we obtain the desired isomorphism $H_\lambda(W, V) \cong H^{n-\lambda}(W, V')$.

<u>Theorem 7.6</u> <u>Basis Theorem</u>

Suppose $(W; V, V')$ is a triad of dimension n possessing a Morse function f with all critical points of index λ and on the same level; and let ξ be a gradient-like vector field for f. Assume that $2 \leq \lambda \leq n - 2$ and that W is connected. Then given any basis for $H_\lambda(W, V)$, there exist a Morse function f' and a gradient-like vector field ξ' for f' which agree with f and ξ in a neighborhood of $V \cup V'$ and are such that f' has the same critical points as f, all on the same level, and the left-hand disks for ξ', when suitably oriented, determine the given basis.

<u>Proof</u>: Let $p_1, \ldots, p_k$ be the critical points of f and let $b_1, \ldots, b_k$ be the basis of $H_\lambda(W, V) \cong Z \oplus \ldots \oplus Z$ (k-summands) represented by the left-hand disks $D_L(p_1), \ldots, D_L(p_k)$ with any fixed orientations. Let the normal bundles of the right-hand disks $D_R(p_1), \ldots, D_R(p_k)$ be oriented so that the matrix

$(D_R(p_i) \cdot D_L(p_j))$ of intersection numbers is the identity $k \times k$ matrix.

Consider first any oriented λ-disk D smoothly imbedded in W with $\text{Bd}D \subset V$. D represents an element

$$\alpha_1 b_1 + \ldots + \alpha_k b_k \varepsilon H_\lambda(W, V)$$

for some integers $\alpha_1, \ldots, \alpha_k$; that is, D is homolgous to $\alpha_1 D_L(p_1) + \ldots + \alpha_k D_L(p_k)$. It follows from an easily proved relative version of Lemma 6.3 that, for each $j = 1, \ldots, k$

$$\begin{aligned} D_R(p_j) \cdot D &= D_R(p_j) \cdot [\alpha_1 D_L(p_1) + \ldots + \alpha_k D_L(p_k)] \\ &= \alpha_1 D_R(p_j) \cdot D_L(p_1) + \ldots + \alpha_k D_R(p_j) \cdot D_L(p_k) \\ &= \alpha_j . \end{aligned}$$

Thus D represents the element

$$D_R(p_1) \cdot D \;\; b_1 + \ldots + D_R(p_k) \cdot D \;\; b_k .$$

We shall construct f' and ξ' so that the new oriented left-hand disks are $D_L'(p_1), D_L(p_2), \ldots, D_L(p_k)$ with $D_R(p_1) \cdot D_L'(p_1) = D_R(p_2) \cdot D_L'(p_1) = +1$ and $D_R(p_j) \cdot D_L'(p_1) = 0$ for $j = 3, 4, \ldots, k$. It follows from the previous paragraph that the new basis is then $b_1 + b_2, b_2, \ldots, b_k$. One can also cause a basis element to be replaced by its negative simply by reversing the orientation of the corresponding left-hand disk. Since a composition of such elementary operations yields any desired basis, this will complete the proof.

The steps involved are roughly as follows: increase f in a neighborhood of p_1, alter the vector field so that the left-hand disk of p_1 "sweeps across" p_2 with positive sign, and then readjust the function so that there is only one critical value.

More precisely, using 4.1 find a Morse function f_1 which agrees with f outside a small neighborhood of p_1 such that $f_1(p_1) > f(p_1)$ and f_1 has the same critical points and gradient-like vector field as f. Choose t_0 so that $f_1(p_1) > t_0 > f(p)$ and set $V_0 = f_1^{-1}(t_0)$.

The left-hand $(\lambda - 1)$-sphere S_L of p_1 in V_0 and the right-hand $(n - \lambda - 1)$-spheres $S_R(p_i)$ of the p_i, $2 \leq i \leq k$, lying in V_0 are disjoint. Choose points $a \in S_L$ and $b \in S_R(p_2)$. Since W, and hence V_0, is connected, there is an embedding $\varphi_1 : (0, 3) \longrightarrow V_0$ such that $\varphi_1(0, 3)$ intersects each of S_L and $S_R(p_2)$ once, transversely, in $\varphi_1(1) = a$ and $\varphi_2(2) = b$, and such that $\varphi_1(0, 3) \cap (S_R(p_3) \cup \ldots \cup S_R(p_k)) = \emptyset$.

<u>Lemma 7.7</u> There exists an embedding $\varphi : (0, 3) \times R^{\lambda-1} \times R^{n-\lambda-1} \longrightarrow V_0$ such that

1.) $\varphi(s, 0, 0) = \varphi_1(s)$ for $s \in (0, 3)$,

2.) $\varphi^{-1}(S_L) = 1 \times R^{\lambda-1} \times 0$, $\varphi^{-1}(S_R(p_2)) = 2 \times 0 \times R^{n-\lambda-1}$ and

3.) the image of φ misses the other spheres. Moreover, φ can be chosen so that it maps $1 \times R^{\lambda-1} \times 0$ into S_L with positive orientation and so that $\varphi((0, 3) \times R^{\lambda-1} \times 0)$ intersects $S_R(p_2)$ at $\varphi(2, 0, 0) = b$ with intersection number +1.

<u>Proof</u>:

Choose a Riemann metric for V_0 so that the arc $A = \varphi_1(0, 3)$ is orthogonal to S_L and to $S_R(p_2)$ and so that these spheres are totally geodesic submanifolds of V_0 (cf. Lemma 6.7).

Let $\mu(a)$ and $\mu(b)$ be orthonormal $(\lambda - 1)$-frames at a and b such that $\mu(a)$ is tangent to S_L at a with positive orientation and $\mu(b)$ is orthogonal to $S_R(p_2)$ at b with intersection number +1. The bundle over A of orthonormal $(\lambda-1)$-frames of vectors orthogonal to A is a trivial bundle with fiber the Stiefel manifold $V_{\lambda-1}(R^{n-2})$, which is connected since $\lambda - 1 < n - 2$. Hence we may extend to a smooth cross-section μ along all of A.

The bundle over A of orthonormal $(n-\lambda-1)$-frames of vectors orthogonal to A and to μ is a trivial bundle with fiber $V_{n-\lambda-1}(R^{n-\lambda-1})$. Let η be a smooth cross-section.

Now use the exponential map associated to the metric to define the desired embedding φ with the help of the $(n-2)$-frames $\mu\eta$. The details are similar to those in the completion of the proof of Lemma 6.7, page 83 . This finishes the proof of Lemma 7.7.

Completion of Proof of the Basis Theorem 7.6

Using φ we construct an isotopy of V_0 which sweeps S_L across $S_R(p_2)$, as follows. (See Figure 7.2)

Fix a number $\delta > 0$ and let $\alpha : R \longrightarrow [1, 2\frac{1}{2}]$ be a smooth function such that $\alpha(u) = 1$ for $u \geq 2\delta$ and $\alpha(u) > 2$ for $u \leq \delta$.

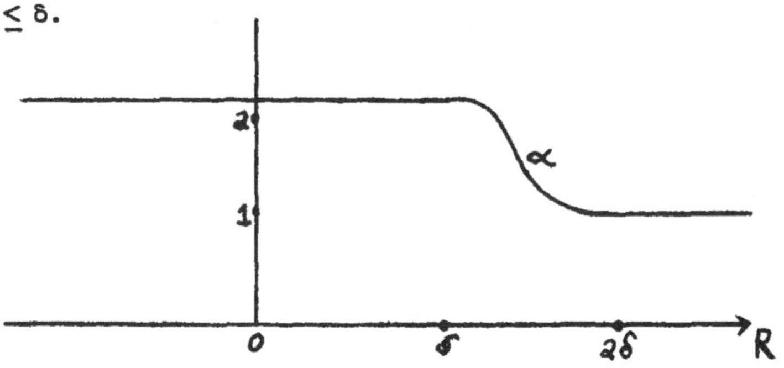

Figure 7.1

As in the last paragraph in the proof of Theorem 6.6, page 74, construct an isotopy H_t of $(0, 3) \times R^{\lambda-1} \times R^{n-\lambda-1}$ such that

1.) H_t is the identity outside some compact set, $0 \leq t \leq 1$.

2.) $H_t(1, \vec{x}, 0) = (t\,\alpha\,(|\vec{x}|^2) + (1 - t), \vec{x}, 0)$ for $\vec{x} \in R^{\lambda-1}$.

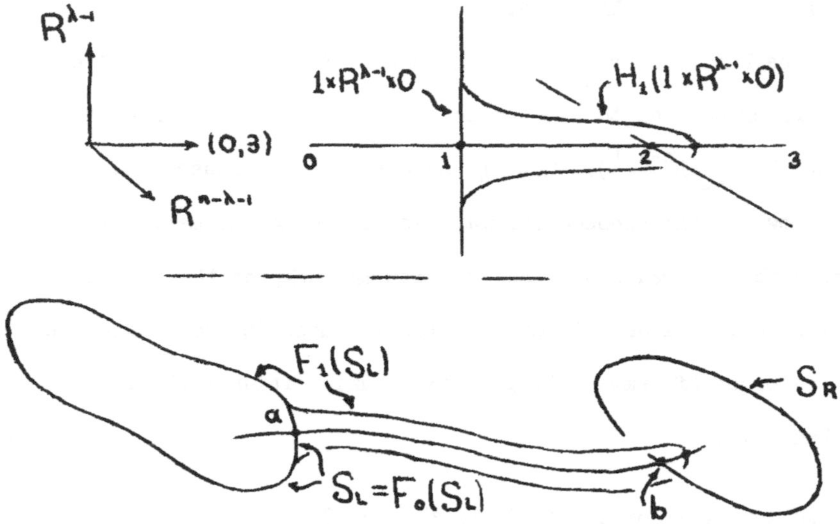

Figure 7.2

Define an isotopy F_t of V_0 by $F_t(v) = \varphi \circ H_t \circ \varphi^{-1}(v)$ for $v \in$ Image (φ) and $F_t(v) = v$ otherwise. From property 1.) of H_t we see that F_t is well-defined.

Now using Lemma 3.5 find a product neighborhood $V_0 \times [0, 1]$ embedded in W on the right side of V_0 such that it contains no critical points and $V_0 \times 0 = V_0$. Using the isotopy F_t, alter the vector field ξ on this neighborhood as in Lemma 4.7, obtaining a new vector field ξ' on W.

Since ξ and ξ' agree to the left of V_0 (that is, on $f_1^{-1}(-\infty, t_0]$), it follows that the right-hand spheres in V_0

associated to ξ' are still $S_R(p_2), \ldots, S_R(p_k)$. The left-hand sphere of p_1 associated to ξ' is $S_L' = F_0(S_L)$. From property 2.) of H_0 we know that S_L' misses $S_R(p_3), \ldots, S_R(p_k)$. Hence by 4.2 we can find a Morse function f' agreeing in a neighborhood of BdW with f_1 (and so with f), having ξ' as associated gradient-like vector field, and having only one critical value.

This completes the construction of f' and ξ'. It remains to show that the new left-hand disks represent the desired basis.

The left-hand disks of $p_2, \ldots, p_k$ associated to ξ' are still $D_L(p_2), \ldots, D_L(p_k)$ since $\xi' = \xi$ to the left of the neighborhood $V_0 \times [0, 1]$, that is, on $f_1^{-1}(-\infty, t_0]$. Since $\xi' = \xi$ also to the right of $V_0 \times [0, 1]$, the new left-hand disk $D_L'(p_1)$ intersects $D_R(p_1)$ at $p_1 = D_L'(p_1) \cap D_R(p_1)$ with intersection number $D_R(p_1) \cdot D_L'(p_1) = +1$. It follows from property 2.) of H_t that $D_L'(p_1)$ intersects $D_R(p_2)$ in a single point, transversely, with intersection number $D_R(p_2) \cdot D_L'(p_1) = +1$. Finally, property 3.) of φ implies that $D_L'(p_1)$ is disjoint from $D_R(p_3), \ldots, D_R(p_k)$ and hence that $D_R(p_i) \cdot D_L'(p_1) = 0$ for $i = 3, \ldots, k$. Thus the basis for $H_\lambda(W, V)$ represented by the left-hand disks associated to ξ' is indeed $b_1 + b_2, b_2, \ldots, b_k$, which completes the proof of 7.6.

<u>Theorem 7.8</u> Suppose $(W; V, V')$ is a triad of dimension $n \geq 6$ possessing a Morse function with no critical points of indices 0, 1 or $n - 1$, n. Futhermore, assume that W, V and V' are all simply connected (hence orientable) and that $H_*(W, V) = 0$. Then $(W; V, V')$ is a product cobordism.

Let c denote the cobordism $(W; V, V')$. It follows from Theorem 4.8 that we can factor $c = c_2 c_3 \ldots c_{n-2}$ so that c admits a Morse function f whose restriction to each c_λ is a Morse function all of whose critical points are on the same level and have index λ. With the notation as in Theorem 7.4 we have the sequence of free abelian groups $C_{n-2} \xrightarrow{\partial} C_{n-3} \xrightarrow{\partial} \ldots \xrightarrow{\partial} C_{\lambda+1} \xrightarrow{\partial} C_\lambda \xrightarrow{\partial} \ldots \xrightarrow{\partial} C_2$. For each λ, choose a basis $z_1^{\lambda+1}, \ldots, z_{k_{\lambda+1}}^{\lambda+1}$ for the kernel of $\partial : C_{\lambda+1} \longrightarrow C_\lambda$. Since

$H_*(W, V) = 0$ it follows from Theorem 7.4 that the above sequence is exact and hence that we may choose $b_1^{\lambda+1}, \ldots, b_{k_\lambda}^{\lambda+1} \in C_{\lambda+1}$ such that $b_i^{\lambda+1} \xrightarrow{\partial} z_i^\lambda$ for $i = 1, \ldots, k_\lambda$. Then $z_1^{\lambda+1}, \ldots, z_{k_{\lambda+1}}^{\lambda+1}, b_1^{\lambda+1}, \ldots, b_{k_\lambda}^{\lambda+1}$ is a basis for $C_{\lambda+1}$.

Since $2 \leq \lambda \leq \lambda + 1 \leq n - 2$, using Theorem 7.6 we can find a Morse function f' and gradient-like vector field ξ' on c so that the left-hand disks of c_λ and $c_{\lambda+1}$ represent the chosen bases for C_λ and $C_{\lambda+1}$.

Let p and q be the critical points in c_λ and $c_{\lambda+1}$ corresponding to z_1^λ and $b_1^{\lambda+1}$. By increasing f' in a neighborhood of p and decreasing f' in a neighborhood of q (see 4.1, 4.2) we obtain $c_\lambda c_{\lambda+1} = c'_\lambda c_p c_q c'_{\lambda+1}$, where c_p has exactly one critical point p and c_q has exactly one critical point q. Let V_0 be the level manifold between c_p and c_q.

It is easy to verify that $c_p c_q$ and its two end manifolds are all simply connected (compare Remark 1, page 70). Since $\partial b_1^{\lambda+1} = z_1^{\lambda}$ the spheres $S_R(p)$ and $S_L(q)$ in V_o have intersection number ± 1. Hence the Second Cancellation Theorem 6.4 or Corollary 6.5 implies that $c_p c_q$ is a product cobordism and that f' and its gradient-like vector field can be altered on the interior of $c_p c_q$ so that f' has no critical points there. Repeating this process as often as possible we clearly eliminate all critical points. Then, in view of Theorem 3.4, the proof of Theorem 7.8 is complete.

§8 Elimination of Critical Points of Index 0 and 1.

Consider a smooth triad $(W^n; V, V')$. We will always assume that it carries a 'self indexing' Morse function f (see 4.9) and an associated gradient-like vector field ξ. Let $W_k = f^{-1}[-\frac{1}{2}, k + \frac{1}{2}]$, $k = 0, 1, \ldots, n$, and $V_{k+} = f^{-1}(k + \frac{1}{2})$.

Theorem 8.1

Index 0) If $H_0(W, V) = 0$, the critical points of index 0 can be cancelled against an equal number of critical points of index 1.

Index 1) Suppose W and V are simply connected and $n \geq 5$. If there are no critical points of index 0 one can insert for each index 1 critical point a pair of auxiliary index 2 and index 3 critical points and cancel the index 1 critical points against the auxiliary index 2 critical points. (Thus one 'trades' the critical points of index 1 for an equal number of critical points of index 3.)

Remark: The method we used to cancel critical points of index $2 \leq \lambda \leq n - 2$ in Theorem 7.8 fails at index 1 for the following reason. The Second Cancellation Theorem 6.4 holds for $\lambda = 1$, $n \geq 6$. (see page 70) , but we would want to apply it where the simple connectivity assumption of 6.4 is spoiled by the presence of several index 1 critical points.

Proof for Index 0:

If in V_{0+} we can always find S_R^{n-1} and S_L^0 intersecting in a single point, then the proof will follow from 4.2, 5.4 (The First Cancellation Theorem) and a finite induction (cf. proof for index 1 below). Consider homology with coefficients in $Z_2 = Z/2Z$. Since $H_o(W, V; Z_2) = 0$, by Theorem 7.4, $H_1(W_1, W_o; Z_2) \xrightarrow{\partial} H(W_o, V; Z_2)$ is onto. But ∂ is clearly given by the matrix of intersection numbers modulo 2 of the right-hand $(n-1)$-spheres and left-hand 0-spheres in V_{o+}. Hence for any S_R^{n-1} there is at least one S_L^o with $S_R^{n-1} \cdot S_L^o \not\equiv 0$ mod 2. This says $S_R^{n-1} \cap S_L^o$ consists of an odd number of points which can only be 1. This completes the proof for index 1.

To construct auxiliary critical points we will need

Lemma 8.2

Given $0 \leq \lambda < n$, there exists a smooth map $f : R^n \longrightarrow R$ so that $f(x_1, \ldots, x_n) = x_1$ outside of a compact set; and so that f has just two critical points p_1, p_2, non-degenerate, of indices $\lambda, \lambda + 1$ respectively with $f(p_1) < f(p_2)$.

Proof: We identify R^n with $R \times R^\lambda \times R^{n-\lambda-1}$; and denote a general point by (x, y, z). Let y^2 be the square of the length of $y \in R^\lambda$.

Choose a function $s(x)$ with compact support so that $x + s(x)$ has two non-degenerate critical points, say x_o, x_1.

Figure 8.1

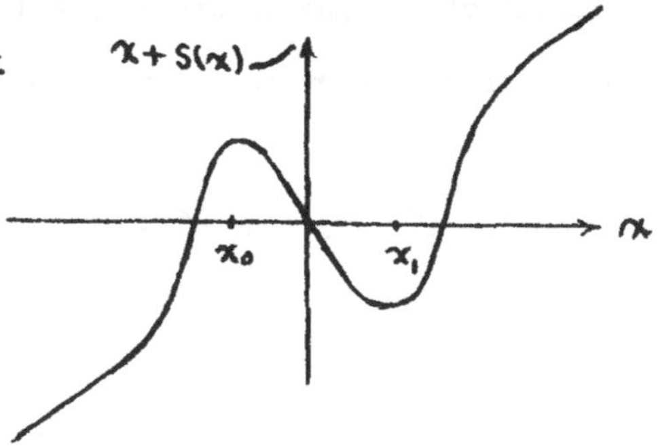

First consider the function $x + s(x) - y^2 + z^2$ on R^n. This has two non-degenerate critical points $(x_0, 0, 0)$ and $(x_1, 0, 0)$ with the correct indices.

Now "taper" this function off as follows. Choose three smooth functions $\alpha, \beta, \gamma : R \longrightarrow R_+$ with compact support so that

1) $\alpha(t) = 1$ for $|t| \leq 1$

2) $|\alpha'(t)| < 1/\underset{x}{\text{Max}} |s(x)|$ for all t (Primes denote derivatives.)

3) $\beta(t) = 1$ whenever $\alpha(t) \neq 0$.

4) $\gamma(x) = 1$ whenever $s'(x) \neq 0$.

5) $|\gamma'(x)| < 1/\underset{t}{\text{Max}} (t\, \beta(t))$

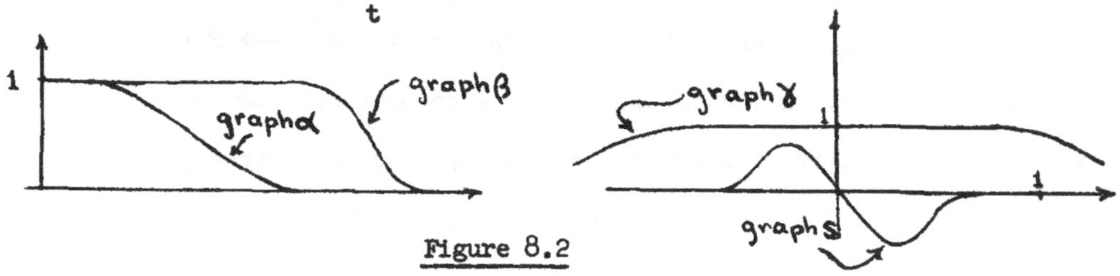

Figure 8.2

Now let

$$f = x + s(x)\, \alpha(y^2 + z^2) + \gamma(x)(-y^2 + z^2)\, \beta(y^2 + z^2).$$

Note that

(a) $f - x$ has compact support

(b) Within the interior of the region where $\alpha = 1$ (hence $\beta = 1$) and $\gamma = 1$ this is our old function, with the old critical points.

(c) $\dfrac{\partial f}{\partial x} = 1 + s'(x)\, \alpha(y^2 + z^2) + \gamma'(x)(-y^2 + z^2)\, \beta(y^2 + z^2)$

The third term has absolute value < 1 by (5). Hence if $s'(x) = 0$ or $\alpha(y^2 + z^2) = 0$ we have $\dfrac{\partial f}{\partial x} \neq 0$. Thus we must only look at the region where $s'(x) \neq 0$ (hence $\gamma = 1$) and

$\alpha(y^2 + z^2) \neq 0$ (hence $\beta = 1$) to look for critical points.

(d) Within the region $\gamma = 1$, $\beta = 1$ we have grad (f) = $(1 + s'(x)\,\alpha(y^2 + z^2),\ 2y(s(x)\,\alpha'(y^2 + z^2) - 1),\ 2z(s(x)\,\alpha'(y^2 + z^2) + 1))$. But $s(x)\,\alpha'(y^2 + z^2) \pm 1 \neq 0$ by (2). Hence the gradient can vanish only when $y = 0$, $z = 0$, and therefore $\alpha = 1$. But this case has already been described in (b).

Proof 8.1 for Index 1:

The given situation may be represented schematically

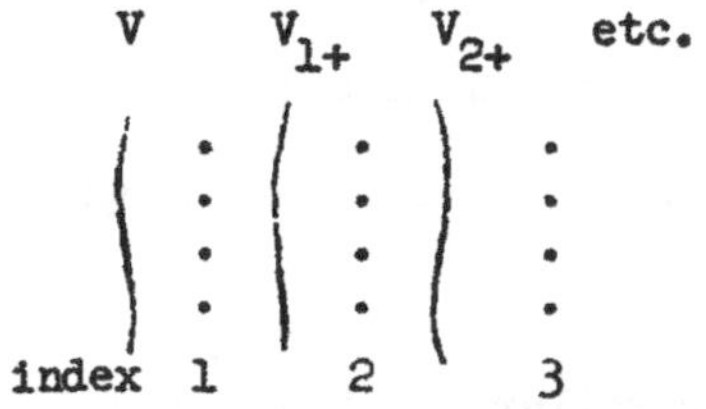

The first step of the proof is to construct, for any right-hand $(n - 2)$-sphere in V_{1+} of a critical point p, a suitable 1-sphere to be the left-hand sphere of the index 2 critical point that will cancel p.

Lemma 8.3

If S_R^{n-2} is a right-hand sphere in V_{1+}, there exists a 1-sphere imbedded in V_{1+} that has one transverse intersection with S_R^{n-2} and meets no other right-hand sphere.

Proof:

Certainly there exists a small imbedded 1-disc $D \subset V_{1+}$, which, at its midpoint q_0, transversely intersects S_R^{n-2}, and which has no other intersection with right-hand spheres. Translate the end points of D left along the trajectories of ξ to a pair of points in V. Since V is connected, and of dimension $n - 1 \geq 2$, these points may be joined by a smooth path in V which avoids the left hand 0-spheres in V. This path may be translated back to a smooth path that joins the end points of D in V_{1+} and avoids all right-hand spheres. Now one can easily construct a smooth map $g : S^1 \longrightarrow V_{1+}$ such that

(a) $g^{-1}(q_0)$ is a point $a \in S^1$ and g smoothly imbeds a closed neighborhood A of a onto a neighborhood of q_0 in D.

(b) $g(S^1 - a)$ meets no right-hand $(n - 2)$-sphere.

Since $\dim V = n - 1 \geq 3$, Whitney's theorem 6.12 provides a smooth imbedding with these properties. This completes the proof of 8.3.

We will need the following corollary of Theorems 6.11, 6.12.

<u>Theorem 8.4</u>

If two smooth imbeddings of a smooth manifold M^m into a smooth manifold N^n are homotopic, then they are smoothly isotopic provided $n \geq 2m + 3$.

<u>Remark</u>: Actually 8.4 holds with $n \geq 2m + 2$ (see Whitney [16])

<u>Proof of Theorem 8.1 for Index 1 continued</u>:

Notice that V_{2+} is always simply connected. In fact the inclusion $V_{2+} \subset W$ factors into a sequence of inclusions that are alternately inclusions associated with cell attachments and inclusions that are homotopy equivalences. (see 3.14). The cells attached are of dimension $n - 2$ and $n - 1$ going to the left and of dimension 3, 4, ... going to the right. Thus V_{2+} is connected since W is, and $\pi_1(V_{2+}) = \pi_1(W) = 1$ (cf. Remark 1 page 70). Given any critical point p of index 1, we construct an 'ideal' 1-sphere S in V_{1+} as in Lemma 8.3. After adjusting ξ if necessary to the right of V_{2+} we may assume that S meets no left hand 1-spheres in V_{1+}. (see 4.6, 4.7). Then we can translate S right to a 1-sphere S_1 in V_{2+}.

In a collar neighborhood extending to the right of V_{2+}, we can choose co-ordinate functions $x_1, \ldots, x_n$ embedding an open set U into R^n so that $f|U = x_n$ (cf. proof of 2.9). Use Lemma 8.2 to alter f on a compact subset of U inserting a pair q, r, with $f(q) < f(r)$, of 'auxiliary' critical points of index 2 and 3. (see Figure 8.3).

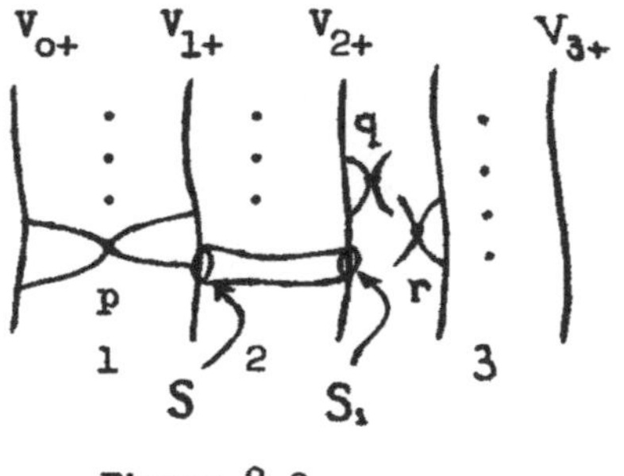

Figure 8.3

Let S_2 be the left-hand 1-sphere of q in V_{2+}. Since V_{2+} is simply connected, 8.4 and 5.8 imply that there is an isotopy of the identity $V_{2+} \longrightarrow V_{2+}$ that carries S_2 to S_1. Thus after an adjustment of ξ to the right of V_{2+} (see 4.7), the left-hand sphere of q in V_{2+} will be S_1. Then the left-hand sphere of q in V_{1+} is S, which, by construction intersects the right-hand sphere of p in a single point, transversely.

Without changing ξ, alter f (by 4.2) on the interiors of $f^{-1}[\frac{1}{2}, 1\frac{1}{2}]$ and of $f^{-1}[1\frac{1}{2}, k]$, $k = (f(q) + f(r))/2$, increasing the level of p and lowering the level of q so that for some $\delta > 0$

$$1 + \delta < f(p) < 1\frac{1}{2} < f(q) < 2 - \delta$$

Now use the First Cancellation Theorem to alter f and ξ on $f^{-1}[1 + \delta, 2 - \delta]$ eliminating the two critical points p and q. Finally move the critical level of r right to 3 (using 4.2).

We have now 'traded' p for r, and the process may be repeated until no critical points of index 1 remain. This completes the proof of Theorem 8.1.

§9. The h-Cobordism Theorem and Some Applications.

Here is the theorem we have been striving to prove.

Theorem 9.1 The h-Cobordism Theorem

Suppose the triad $(W^n; V, V')$ has the properties

1) W, V and V' are simply connected.

2) $H_*(W, V) = 0$

3) $\dim W = n \geq 6$

Then W is diffeomorphic to $V \times [0, 1]$

Remark: The condition 2) is equivalent to 2)' $H_*(W, V') = 0$. For $H_*(W, V) = 0$ implies $H^*(W, V') = 0$ by Poincaré duality. But $H^*(W, V') = 0$ implies $H_*(W, V') = 0$. Similarly 2)' implies 2).

Proof: Choose a self-indexing Morse function f for $(W; V, V')$. Theorem 8.1 provides for the elimination of critical points of index 0 and 1. If we replace the Morse function f by $-f$ the triad is 'turned about' and critical points of index λ become critical points of index $n - \lambda$. Thus critical points of (original) index n and $n - 1$ may also be eliminated. Now Theorem 7.8 gives the desired conclusion.

Definition 9.2 A triad $(W; V, V') = 0$ is an h-cobordism and V is said to be h-cobordant to V' if both V and V' are deformation retracts of W.

Remark: It is an interesting fact (which we will not use) that an equivalent version of Theorem 9.1 is obtained if we substitute for 2) the apparently stronger condition that $(W; V, V')$ be an h-cobordism. Actually 1) and 2) together imply that $(W; V, V')$ is an h-cobordism. In fact

(i) $\pi_1(V) = 0$, $\pi_1(W, V) = 0$, $H_*(W, V) = 0$ together imply

(ii) $\pi_i(W, V) = 0 \quad i = 0, 1, 2, \ldots$

by the (relative) Hurewicz isomorphism theorem (Hu, [20, p.166]; Hilton [21, p.103]). In view of the fact that (W, V) is a triangulable pair (Munkres [5, p.101]) (ii) implies that a strong deformation retraction $W \longrightarrow V$ can be constructed. (See Hilton [21, p.98 Thm 1.7].) Since 2) implies $H_*(W, V') = 0$, V' is, by the same argument, a (strong) deformation retract of W.

An important corollary of Theorem 9.1 is

Theorem 9.2

Two simply connected closed smooth manifolds of dimension ≥ 5 that are h-cobordant are diffeomorphic.

A Few Applications (see also Smale [22] [6])

Proposition A) Characterizations of the smooth n-disc D^n, $n \geq 6$.

Suppose W^n is a compact simply connected smooth n-manifold, $n \geq 6$, with a simply connected boundary. Then the following four assertions are equivalent.

1). W^n is diffeomorphic to D^n.

2). W^n is homeomorphic to D^n.

3). W^n is contractible.

4). W^n has the (integral) homology of a point.

Proof: Clearly 1) => 2) => 3) => 4). So we prove 4) => 1). If D_o is a smooth n-disc imbedded in Int W, then (W - Int D_o; BdD_o, V) satisfies the conditions of the h-Cobordism Theorem. In particular, (by excision) $H_*(W - \text{Int } D_o, BdD_o) \cong H_*(W, D_o) = 0$.

Consequently the cobordism $(W^n; \emptyset, V)$ is a composition of $(D_o; \emptyset, BdD_o)$ with a product cobordism (W - Int D_o; BdD_o, V). It follows from 1.4 that W is diffeomorphic to D_o.

Proposition B) The Generalized Poincaré Conjecture in dimensions ≥ 5. (See Smale [21].)

If M^n, $n \geq 5$, is a closed simply-connected smooth manifold with the (integral) homology of the n-sphere S^n, then M^n is homeomorphic to S^n. If $n = 5$ or 6, M^n is diffeomorphic to S^n.

Corollary If a closed smooth manifold M^n, $n \geq 5$, is a homotopy n-sphere (i.e. is of the homotopy type of S^n) then M^n is homeomorphic to S^n.

Remark: There exist smooth 7-manifolds M^7 that are homeomorphic to S^7 but are not diffeomorphic to S^7. (See Milnor [24].)

Proof of B Suppose first that $n \geq 6$. If $D_o \subset M$ is a smooth n-disc, M - Int D_o satisfies the conditions of A).
In particular

$$H_i(M - \text{Int } D_o) \cong H^{n-i}(M - \text{int } D_o, \partial D_o) \text{ (Poincare duality 7.5)}$$
$$\cong H^{n-i}(M, D_o) \text{ (excision)}$$
$$\cong \begin{cases} 0 & \text{if } i > 0 \\ Z & \text{if } i = 0 \end{cases} \text{ (exact sequence)}$$

Consequently $M = (M - \text{Int } D_0) \cup D_0$ is diffeomorphic to a union of two copies D_1^n, D_2^n of the n-disc with the boundaries identified under a diffeomorphism $h : BdD_1^n \longrightarrow BdD_2^n$.

Remark: Such a manifold is called a twisted sphere. Clearly every twisted sphere is a closed manifold with Morse number 2, and conversely.

The proof is completed by showing that any twisted sphere $M = D_1^n \cup_h D_2^n$ is homeomorphic to S^n. Let $g_1 : D_1^n \longrightarrow S^n$ be an imbedding onto the southern hemisphere of $S^n \subset R^{n+1}$ i.e. the set $\{\vec{x} \mid |\vec{x}| = 1, x_{n+1} \leq 0\}$. Each point of D_2^n may be written tv, $0 \leq t \leq 1$, $v \in BdD_2$. Define $g : M \longrightarrow S^n$ by

(i) $g(u) = g_1(u)$ if $u \in D_1^n$

(ii) $g(tv) = \sin \frac{\pi t}{2} g_1(h^{-1}(v)) + \cos \frac{\pi t}{2} e_{n+1}$ where $e_{n+1} = (0, \ldots, 0, 1) \in R^{n+1}$, for all points tv in D_2^n.

Then g is a well defined 1-1 continuous map onto S^n, and hence is a homeomorphism. This completes the proof for $n \geq 6$. If $n = 5$ we use:

Theorem 9.1 (Kervaire and Milnor [25], Wall [26])

Suppose M^n is a closed, simply connected, smooth manifold with the homology of the n-sphere S^n. Then if $n = 4$, 5, or 6, M^n bounds a smooth, compact, contractible manifold.

Then A) implies that for $n = 5$ or 6 M^n is actually diffeomorphic to S^n.

Proposition C Characterization of the 5-disc

Suppose W^5 is a compact simply connected smooth manifold that has the (integral) homology of a point. Let $V = BdW$.

1) If V is diffeomorphic to S^4 then W is diffeomorphic to D^5.

2) If V is homeomorphic to S^4 then W is homeomorphic to D^5.

<u>Proof of 1)</u> Form a smooth 5-manifold $M = W \cup_h D^5$ where h is a diffeomorphism $V \longrightarrow \mathrm{Bd}D^5 = S^4$. Then M is a simply connected manifold with the homology of a sphere. In B) we proved that M is actually diffeomorphic to S^5. Now we use

<u>Theorem 9.6</u> (Palais [27], Cerf [28], Milnor [12, p.11]) Any two smooth orientation-preserving imbeddings of an n-disc into a connected oriented n-manifold are ambient isotopic.

Thus there is a diffeomorphism $g : M \longrightarrow M$ that maps $D^5 \subset M$ onto a disc D_1^5 such that $D_2^5 = M - \mathrm{Int}\, D_1^5$ is also a disc. Then g maps $W \subset M$ diffeomorphically onto D_2^5.

<u>Proof of 2)</u> Consider the double $D(W)$ of W (i.e. two copies of W with the boundaries identified — see Munkres [5, p.54]). The submanifold $V \subset D(W)$ has a bicollar neighborhood in $D(W)$, and $D(W)$ is homeomorphic to S^5 by B). Brown [23] has proved:

<u>Theorem 9.7</u> If an $(n - 1)$-sphere Σ, topologically imbedded in S^n, has a bicollar neighborhood, then there exists a homeomorphism $h : S^n \longrightarrow S^n$ that maps Σ onto $S^{n-1} \subset S^n$. Thus $S^n - \Sigma$ has two components and the closure of each is an n-disc with boundary Σ.

It follows that W is homeomorphic to D^5. This completes the proof of C).

<u>Proposition D) The Differentiable Schoenfliess Theorem in Dimensions ≥ 5.</u>

Suppose Σ is a smoothly imbedded $(n-1)$-sphere in S^n. If $n \geq 5$, there is a smooth ambient isotopy that carries Σ onto the equator $S^{n-1} \subset S^n$.

<u>Proof:</u> $S^n - \Sigma$ has two components (by Alexander duality) and hence 3.6 shows that Σ is bicollared in S^n. The closure in S^n of a component of $S^n - \Sigma$ is a smooth simply connected manifold D_o with boundary Σ and with the (integral) homology of a point. For $n \geq 5$, D_o is actually diffeomorphic to D^n by A) and C). Then the theorem of Palais and Cerf (9.6) provides an ambient isotopy that carries D_o to the lower hemisphere and hence $BdD_o = \Sigma$ to the equator.

<u>Remark:</u> This shows that if $f : S^{n-1} \longrightarrow S^n$ is a smooth imbedding, then f is smoothly isotopic to a map onto $S^{n-1} \subset S^n$; but it is not in general true that f is smoothly isotopic to the inclusion $i : S^{n-1} \longrightarrow S^n$. It is false if $f = i \circ g$, where $g : S^{n-1} \longrightarrow S^{n-1}$ is a diffeomorphism which does not extend to a diffeomorphism $D^n \longrightarrow D^n$. (The reader can easily show that g extends to D^n if and only if the twisted sphere $D_1^n \cup_g D_2^n$ is diffeomorphic to S^n.) In fact if f is smoothly isotopic to i, by the Isotopy Extension Theorem 5.8, there exists a diffeomorphism $d : S^n \longrightarrow S^n$ such that $d \circ i = f = i \circ g$. This gives two extensions of g to a diffeomorphism $D^n \longrightarrow D^n$.

Concluding Remarks:

It is an open question whether the h-Cobordism Theorem is true for dimensions $n < 6$. Let $(W^n; V, V')$ be an h-cobordism where W^n is simply connected and $n < 6$.

$n = 0, 1, 2$: The theorem is trivial (or vacuous).

$n = 3$: V and V' must be 2-spheres. Then the theorem is easily deduced from the classical Poincaré Conjecture: Every compact smooth 3-manifold which is homotopy equivalent to S^3 is diffeomorphic to S^3. Since every twisted 3-sphere (see page 110) is diffeomorphic to S^3 (see Smale [30], Munkres [31]) the theorem is actually equivalent to this conjecture.

$n = 4$: If the classical Poincaré Conjecture is true V and V' must be 3-spheres. Then the theorem is readily seen to be equivalent to the '4-Disk Conjecture': Every compact contractible smooth 4-manifold with boundary S^3 is diffeomorphic to D^4. Now a difficult theorem of Cerf [29] says that every twisted 4-sphere is diffeomorphic to S^4. It follows that this conjecture is equivalent to: Every compact smooth 4-manifold which is homotopy equivalent to S^4 is diffeomorphic to S^4.

$n = 5$: Proposition C) implies that the theorem does hold when V and V' are diffeomorphic to S^4. However there exist many types of closed simply connected 4-manifolds. Barden (unpublished) showed that if there exists a diffeomorphism $f : V' \longrightarrow V$ homotopic to $r|_{V'}$ where $r : W \longrightarrow V$ is a deformation retraction, then W is diffeomorphic to $V \times [0, 1]$. (See Wall [38], also [37].)

REFERENCES

[1] de Rham, Variétés Différentiables, Hermann, Paris, 1960.

[2] S. Eilenberg and N. Steenrod, Foundations of Algebraic Topology, Princeton University Press, Princetnn, N. J., 1952.

[3] S. Lang, Introduction to Differentiable Manifolds, Interscience, New York, N. Y., 1962.

[4] J. Milnor, Morse Theory, Princeton University Press, Princeton, N. J., 1963.

[5] J. Munkres, Elementary Differential Topology, Princeton University Press, Princeton, N. J., 1963.

[6] S. Smale, On the Structure of Manifolds, Amer. J. of Math. vol. 84 (1962), pp. 387-399.

[7] H. Whitney, The Self-intersections of a Smooth n-manifold in 2n-space, Annals of Math. vol. 45 (1944), pp. 220-246.

[8] S. Smale, On Gradient Dynamical Systems, Annals of Math. vol. 74 (1961), pp. 199-206.

[9] A. Wallace, Modifications and Cobounding Manifolds, Canadian J. Math. 12 (1960), 503-528.

[10] J. Milnor, The Sard-Brown Theorem and Elementary Topology, Mimeographed Princeton University, 1964.

[11] M. Morse, Differential and Combinatorial Topology, (Proceedings of a symposium in honour of Marston Morse) Princeton University Press (to appear).

[12] J. Milnor, Differential Structures, lecture notes mimeographed, Princeton University, 1961.

[13] R. Thom, La classification des immersions, Séminaire Bourbaki, 1957.

[14] J. Milnor, Characteristic Classes, notes by J. Stasheff, mimeographed Princeton University, 1957.

[15] J. Milnor, Differential Topology, notes by J. Munkres, mimeographed Princeton University, 1958.

[16] H. Whitney, Differentiable Manifolds, Annals of Math. vol. 37 (1936), pp. 645-680.

[17] R. Crowell and R. Fox, Introduction to Knot Theory, Ginn and Co., 1963.

[18] N. Steenrod, Topology of Fibre Bundles, Princeton University Press, 1951.

[19] J. Milnor, Characteristic Classes Notes, Appendix A, mimeographed, Princeton University, March, 1964.

[20] S. Hu, Homotopy Theory, Academic Press, 1959.

[21] P. Hilton, An Introduction to Homotopy Theory, Cambridge Tracts in Mathematics and Mathematical Physics, No. 43 Cambridge University Press, 1961.

[22] S. Smale, Generalized Poincaré's Conjecture in Dimensions Greater than 4, Annals of Math., vol. 64 (1956), 399-405.

[23] M. Brown, A Proof of the Generalized Shoenfliess Theorem, Bull. AMS, 66 (1960), pp. 74-76.

[24] J. Milnor, On Manifolds Homeomorphic to the 7-Sphere, Annals of Math, vol. 64, No. 2, (1956), pp. 399-405.

[25] M. Kervaire and J. Milnor, Groups of Homotopy Spheres, Annals of Math., vol. 77, No. 3, (1963), pp. 504-537.

[26] C. T. C. Wall, Killing the Middle Homotopy Groups of Odd Dimensional Manifolds, Trans. Amer. Math. Soc. 103, (1962), pp. 421-433.

[27] R. Palais, Extending Diffeomorphisms, Proc. Amer. Math. Soc. 11 (1960), 274-277.

[28] J. Cerf, Topologie de certains espaces de plongements, Bull. Soc. Math. France, 89 (1961), pp. 227-380.

[29] J. Cerf, La nullité du groupe Γ_4, Sém. H. Cartan, Paris 1962/63, Nos. 8, 9, 10, 20, 21.

[30] S. Smale, Diffeomorphisms of the 2-sphere, Proc. Amer. Math. Soc. 10 (1959), 621-626.

[31] J. Munkres, Differential Isotopies on the 2-sphere, Michigan Math. J. 7 (1960), 193-197.

[32] W. Huebsch and M. Morse, The Bowl Theorem and a Model Nondegenerate Function, Proc. Nat. Acad. Sci. U. S. A., vol. 51 (1964), pp. 49-51.

[33] D. Barden, Structure of Manifolds, Thesis, Cambridge University, 1963.

[34] J. Milnor, Two Complexes Which Are Homeomorphic But Combinatorially Distinct, Annals of Math. vol. 74 (1961), pp. 575-590.

[35] B. Mazur, Differential Topology from the Point of View of Simple Homotopy Theory, Publications Mathematiques, Institute des Hautes Etudes Scientifiques, No. 15.

[36] C. T. C. Wall, Differential Topology, Cambridge University mimeographed notes, Part IV, 1962.

[37] C. T. C. Wall, On Simply-Connected 4-Manifolds, Journal London Math. Soc., 39 (1964), pp. 141-149.

[38] C. T. C. Wall, Topology of Smooth Manifolds, Journal London Math. Soc., 40 (1965), pp. 1-20.

[39] J. Cerf, Isotopy and Pseudo-Isotopy, I, mimeographed at Cambridge University, 1964.

Printed and bound by CPI Group (UK) Ltd, Croydon, CR0 4YY

27/10/2024

14580235-0001